AIChEMI MODULAR INSTRUCTION

Series C: TRANSPORT

Volume 6: Transport Phenomena—Special Topics

R. J. Gordon, Series Editor

AMERICAN INSTITUTE OF CHEMICAL ENGINEERS

American Institute of Chemical Engineers
345 East 47th Street, New York, N.Y. 10017

ISSN 0270-7632/86/$5.00

ISBN 0-8169-0397-2

CONTENTS

Solutions to the Homework Problems are available as a separate reprint from the AIChE Educational Services Dept., 345 East 46th St., New York, NY 10017. The cost is $5.00.

INTRODUCTION

In 1975 a new venture in education by and for the chemical engineering community was initiated. Prepared by the CACHE Corporation (Computer Aids for Chemical Engineering Education) and under the sponsorship of the National Science Foundation (Grant HES 75-03911), a series of small self-study fundamental concept modules for various areas of chemical engineering were commissioned, Chemical Engineering Modular Instruction, CHEMI.

It has been found in recent studies that modular study is more effective than traditional instruction in both university and continuing education settings. This is due in large measure to the discrete focus of each module, which allows the student to tailor the speed and order of his or her study. In addition, since the modules have different authors, each writing in his or her area of special expertise, they can be produced more quickly, and students may be asured of timely information. Finally, these modules have been tested in the classroom prior to their publication.

The educational effect of modular study is to reduce, in general, the number of hours required to teach a given subject; it is expected that the decreased time and expense involved in engineering education, when aided by modular instruction, will attract a larger number of students to engineering, including those who have not traditionally chosen engineering. For the practicing engineer, the modules are intended to enhance or broaden the skills he or she has already acquired, and to make available new fields of expertise.

The modules were designed with a variety of applications in mind. They may be pursued in a number of contexts: as outside study, special projects, entire university courses (credit or non-credit), review courses, or correspondence courses; and they may be studied in a variety of modes: as supplements to course work, as independent study, in continuing education programs, and in the traditional student/teacher mode.

A module was defined as a self-contained set of learning materials tat covers one or more topics. It should be sufficiently detailed that an outside evaluation could identify its educational objectives and determine a student's achievement of these objectives. A module should have the educational equivalent of a one to three hour lecture.

The CHEMI Project Staff included:

E. J. Henley, University of Houston, Director
W. Heenan, Texas A & I University, Assistant Director
Steering Committee:
L. B. Evans, Massachusetts Institute of Technology
G. J. Powers, Carnegie-Mellon University
E. J. Henley, University of Houston
D. M. Himmelblau, University of Texas at Austin
D. A. Mellichamp, University of California at Santa Barbara
R. E. C. Weaver, Tulane University
Editors:
Process Control: T. F. Edgar, University of Texas at Austin
Stagewise and Mass Transfer Operations: E. J. Henley, University of Houston, J. M. Calo, Brown University
Transport: R. J. Gordon, University of Flordia
Thermodynamics: B. M. Goodwin, Northeastern University
Kinetics: B. L. Crynes, Oklahoma State University
H. S. Fogler, University of Michigan
Material and Energy Balances: D. M. Himmelblau, University of Texas at Austin

Curriculum Analysis: E. J. Henley, University of Houston

The second phase of the project, designed to fill in gaps as well as develop new modules, is under the direction of D. M. Himmelblau, University of Texas at Austin.

Steering Committee:

B. Carnahan, University of Michigan
D. E. Griffith, Oklahoma State University
L. Harrisberger, University of Alabama
D. M. Himmelblau, University of Texas at Austin
V. Slamecka, Georgia Institute of Technology
R. Tinker, Technology Education Research Center

Editors (indicates a new task force head)*:

Process Control: T. F. Edgar, University of Texas at Austin
Stagewise and Mass Transfer Operations: J. M. Calo, Brown University
E. J. Henley, University of Houston
Transport: R. J. Gordon, University of Florida
Thermodynamics: G. A. Mansoori*, University of Ilinois at Chicago Circle
Kinetics: B. L. Crynes, Oklahoma State University
H. S. Fogler, University of Michigan
Material and Energy Balances: E. H. Snider*, University of Tulsa
Design of Equipment: J. R. Beckman, Arizona State University

Volume 1 of each series will appear in 1980; Volume 2 in 1981; and so forth. A tentative outline of all volumes to be produced in this series follows:

SERIES C: TRANSPORT

Volume 1. Momentum Transport and Fluid Flow

C1.1	Simplified One-Dimensional Momentum Transport Problems	G. K. Patterson
C1.2	Friction Factor	R. J. Gordon and N. H. Chen
C1.3	Applications of the Steady-State Mechanical Energy Balance	D. W. Hubbard
C1.4	Flow Meters	W. F. Beckwith
C1.5	Packed Beds and Fluidization	W. J. Hatcher, Jr.
C1.6	Multi-Phase Flow	R. A. Greenkorn and D. P. Kessler

Volume 2. Momentum Transport, Viscoelasticity and Turbulence

C2.1	Non-Newtonian Flow I—Characterization of Fluid Behavior	D. V. Boger and A. L. Halmos
C2.2	Non-Newtonian Flow II—Fully Developed Tube Flow	D. V. Boger and A. L. Halmos
C2.3	Viscoelastic Fluid Flow Phenomena	D. V. Boger and R. I. Tanner
C2.4	Turbulence: General Aspects Illustrated by Channel or Pipe Flow	N. S. Berman
C2.5	Turbulent Drag Reduction	G. K. Patterson

Volume 3. Equation of Motion, Boundary Layer Theory and Measurement Techniques

C3.1	Measurements of Local Fluid Velocities	N. S. Berman and H. Usui
C3.2	Equation of Motion	G. K. Patterson
C3.3	Navier Stokes Equation for Steady One-Directional Flow	G. C. April
C3.4	Boundary Layer Theory	R. J. Gordon
C3.5	Boundary Layer Theory: Approximate Solution Techniques	R. L. Cerro
C3.6	Diffusivity Measurement Techniques in Liquids	V. L. Vilker

Volume 4. Mathematical Techniques and Energy Transport

C4.1	Mathematical Techniques I—Separation of Variables	R. S. Subramanian
C4.2	Mathematical Techniques II—Combination of Variables	R. S. Subramanian
C4.3	Elementary Steady-State Heat Conduction	W. J. Hatcher

Volume 5. Mass Transport

Volume 6. Transport Phenomena—Special Topics

Publication and dissemination of these modules is under the direction of Harold I. Abramson, Staff Director, Educational Activities, AIChE. Technical Editor is Lori S. Roth. Chemical engineers in industry or academia who are interested in submitting modules for publication should direct them to H. I. Abramson, Staff Director, Educational Activities, American Institute of Chemical Engineers, 345 East 47th Street, New York, N.Y. 10017.

Bubble Dynamics: An Illustration of Dynamically Coupled Rate Processes

T. G. Theofanous

University of California
Santa Barbara, CA

OBJECTIVES

At the completion of this module, the student should be able to:

1. Estimate vapor bubble growth rates in uniformly superheated liquids.
2. Identify the physical mechanism(s) that control the bubble growth process and those that, although present, can be neglected for the purposes of the calculation, for any given particular case of interest.
3. Evaluate the (heterogeneous and homogeneous) bubble nucleation characteristics.

PREREQUISITE MATHEMATICAL SKILLS

1. Elementary differential equations.

PREREQUISITE ENGINEERING AND SCIENCE SKILLS

1. Equations of change.
2. Elementary thermodynamics of two-phase media.
3. Transient conduction.
4. Dimensional analysis.

INTRODUCTION

The primary purpose of this unit is to discuss the methodology for the study of dynamically *coupled* transport (rate) processes. The subject of "Bubble Dynamics" was selected as the vehicle for this study, not only because it provides a uniquely well suited area of application, but also because it will provide the opportunity to establish the fundamentals underlying the behavior of a very large number of naturally occurring and technologically important processes, not the least of which is the process of boiling.

In the majority of problems, the heat and mass transport rates depend upon the fluid mechanics, however, this dependence has been largely one-sided. That is, the problem could be solved in sequence, starting with the determination of the flow field and proceeding with the solution for the heat and mass transfer rates. An important exception readers may already be familiar with is in the area of natural (free) convection. Here, temperature and concentration gradients give rise to body forces (buoyancy) that produce fluid motion. This motion, in turn, interacts with and affects the development of those gradients, so that a simultaneous solution (coupling) for both processes becomes necessary.

When the degree of coupling is a function of time, one may talk about dynamically coupled processes. Many examples of such processes may be found in the field of gas dynamics. The area of bubble dynamics, which is perhaps closer to the interests and practice of chemical engineering, provides numerous opportunities to investigate such processes. For the purposes of this presentation, the problem of vapor bubble growth has been selected for consideration. In very simple terms, the key interaction (coupling) here is between the liquid inertia, which must be "overcome" to initiate and develop the liquid motion, and the thermal conduction resistance in the liquid surrounding the vapor bubble, which must be overcome in order to maintain the heat supply equivalent to the latent heat requirements for the phase-change process that accompanies growth.

Having to deal with highly transient problems, it is reasonable to begin with a detailed consideration of the initial conditions. Thus, one learns about nucleation, which marks the inception of the bubble growth process. Then the module will continue with the physical interpretation of the individual transport processes that participate in the bubble growth phenomena and their mathematical formulation, emphasizing the simplifying assumptions incorporated in this formulation. There will then be a discussion on the methods and the nature of the solutions, and to conclude, a brief presentation of some more general situations of practical interest.

VAPOR BUBBLE NUCLEATION

Consider carrying out the following very simple experiment. In a clean test tube place a small quantity of Freon-11 and cover it with a thin layer of glycerol. Secure the tube inside a vacuum flask, such as those used for example by chemists for filtration. Replace the filter assemblies with a stopcock equipped with a vacuum gauge (or even better a mercury manometer) and slowly apply vacuum through a respirator. You will observe that soon after the application of vacuum the tube contents "boil-off." It is very likely that during your first try, this "boil-off" process will be slow enough to see the nucleation of vapor bubble(s), which as they grow expell at least a portion of the tube contents and thus destroy the flask vacuum. Now repeat the experiment with a very carefully cleaned (cleaning solution: sodium chromate + concentrated sulphuric acid) and dried (distilled water, alcohol, and acetone in this order) test tube. You will observe a longer delay time and such a violent "boil-off" that although you will be able to *hear*, you will not be able to *see* the expulsion process itself.

All this behavior can be understood with the help of Figure 1. The initial state of the fluid in the test tube is indicated by point O, which for Freon-11 at room temperature and atmospheric pressure is slightly to the left of the vapor pressure curve indicated by the solid line (i.e., Freon-11 boils at 90°F). Since the liquid would have to be heated to its boiling point, $T^*(P_O)$, in order to achieve saturation, the state O is undersaturated (or subcooled) by the amount $T^*(P_O) - T_O$. As the pressure drops, the fluid state follows the trajectory OAB. Saturation is achieved at A. For pressures below the level P_A, the liquid is found in a metastable state; such states can only be achieved in the absence of any liquid-vapor interfaces which can sustain vaporization. At a system pressure P_B, the system is "superheated" by the amount $T_O - T^*(P_B)$, where $T^*(P_B)$ is the saturation temperature of the liquid at pressure P_B. When an opportunity for vaporization appears (nucleation event) somewhere inside the test tube, the *initial* rapid liquid vaporization is supplying vapor at a pressure P_A. Since the flask pressure is P_B, the resulting pressure differential will set the contents of the tube in motion. It is clear that this pressure driving

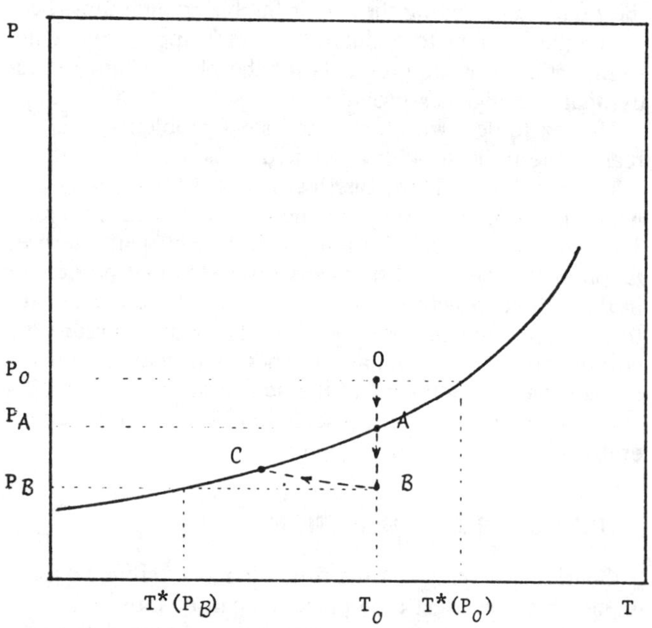

Figure 1. Metastable O liquid states in relation to its vapor pressure curve. Asteriks indicate saturation values, i.e., $p_A = p^*(T_o)$.

force increases with the amount of superheat present and thus, one can see qualitatively the reason for the violence of the event observed experimentally. For a closed system, like our flask, the path BC will be followed spontaneously, terminating at an intermediate state C on the saturation line. Next, let us consider the origin and characteristics of a "nucleation event."

For a system with no free, macroscopic liquid–vapor interfaces (like our system where the free surface of the Freon was "sealed" with a layer of glycerol, which has a much lower vapor pressure), nucleation must initiate from microscopic vapor or gas cavities. Such cavities are present on solid surfaces and may also form spontaneously in the liquid. The crucial aspect concerning the action of a particular cavity as a nucleation site is its size. The size, (radius R), determines the magnitude of the surface tension force, $2\sigma/R$, which is available to balance any difference between the cavity internal pressure (P_i) and ambient liquid pressure P_β. The equilibrium condition may be used to define the critical radius R_{cr}.

$$P_i = \frac{2\sigma}{R_{cr}} + P_\beta \tag{1}$$

It is easy to see that for cavities larger than the critical size, the force imbalance favors growth, and for cavities smaller than the critical size, the force imbalance favors collapse. Thus, any cavities that contain only vapor and are smaller than the critical size will vanish. However, on solid surfaces, especially in the presence, as is always the case, of small amounts of noncondensible gases, vapor and/or gas entrapment in microscopic cavities (surface imperfections) can occur in a fashion that leads to a spectrum of stable, microscopic nuclei. If the radius of curvature of the largest such nuclei is R_m, the maximum possible superheat would correspond to a pressure difference ΔP,

$$\Delta P = \frac{2\sigma}{R_m} \tag{2}$$

For the example of our experiment, if cavity gas pressure is neglected, the minimum possible pressure level P_β would be given by:

$$P_\beta = P_i - \frac{2\sigma}{R_m} \simeq P^*(T_O) - \frac{2\sigma}{R_m} \tag{3}$$

Since there is no direct way to control R_m the nucleation events are somewhat erratic. In a general way, the better the cleaning technique, the smaller is the size of R_m and thus, the larger the vacuum (superheat) achieved in the experiment. Further increases in superheat can be achieved by pressurizing the system, prior to the application of vacuum. This process is effective by forcing the cavities to a smaller size, thus decreasing R_m.

The mode of nucleation originating at pre–existing cavities is the most common one and is called heterogeneous nucleation. In most engineering systems there are enough dissolved gases and other impurities, and surface imperfections, that only very small superheats (a few degrees) become possible. For certain special laboratory conditions and solid/liquid pairs with good wetting properties, heterogeneous nucleation can sustain hundreds of degrees of superheat (for example, for the liquid sodium stainless steel pair, superheats up to 800°F have been measured). However, as long as the metastable liquid has any interfaces with solid surfaces, no matter what precautions have been taken, heterogeneous nucleation cannot be eliminated in practice. This type of nucleation can only be avoided by "sealing" the metastable liquid with another liquid of very low vapor pressure. For example, this can be accomplished by suspending (i.e., by mean of sound waves) a drop of the volatile liquid in a heated oil bath. Such a drop can be superheated well above the heterogeneous nucleation levels. In fact, it will always reach the same, *predictable,* level of superheat, known as the limit of superheat, and it will literally explode upon reaching this limit. This event is called spontaneous nucleation (or homogeneous nucleation) and it is due to thermal fluctuations of molecules in the liquid state, which can randomly achieve configurations such as to produce locally microscopic vapor cavities slightly larger than the "critical size." The rate of formation of such critical cavities, R_{cr}, may be calculated from the molecular thermodynamics of the liquid state and is given by:

$$J = A \ \exp\left\{ -\frac{1\sigma\pi\sigma^3}{3KT_l\Delta P^2} \right\} \tag{4}$$

where: $A \simeq 10^{32}\ \text{cm}^{-3}\ \text{s}^{-1}$ and is known as the frequency factor, K is the Boltzman's constant, T_l is the temperature of the liquid and $\Delta P = P^*(R) - P_l$ is equivalent of the superheat present. This equation, due to its strong dependence on liquid temperature (notice that ΔP also depends strongly on T_l), has the character of a threshold equation. That is, up to the threshold temperature (T_{es}), the nucleation rate is extremely small, i.e., no nucleation is observed experimentally. When the temperature T_{es} is reached, however, and within a very small temperature interval, the nucleation rate increases dramatically, so as to clearly define a limit of superheat. At this limit, the rate of nuclei formation is so high that the drop will literally explode, i.e., the pressurization due to boiling is so great that shock waves are observed.

MATHEMATICAL FORMULATION

Attempting to develop simple mathematical solutions and through them, a first-order understanding of the bubble growth problem, one must identify for study the key physical phenomena that affect bubble growth in a generic fashion. A consideration of qualitative physical behavior may be helpful in that regard.

A vapor-filled critical cavity is depicted in Figure 2a. The cavity is immersed in a superheated liquid with temperature T_∞ and the system pressure is P_∞. The vapor pressure in the cavity corresponds to the saturation pressure at the liquid temperature and since this pressure is balanced exactly by the surface tension force and the ambient pressure, there is no motion (bubble wall velocity is zero, $\dot{R} = 0$). A slight decrease in cavity size will yield monotonic cavity collapse, while a slight increase will yield monotonic growth. Hence the system is at unstable equilibrium. Consider the case of growth. As R increases, the surface tension force decreases and the pressure imbalance across the liquid body increases, leading to its acceleration. On the other hand, bubble growth results in an expansion of its vapor contents and unless an equivalent amount of new vapor is supplied, the bubble internal pressure will decrease. Liquid vaporization, at the bubble wall, can provide the source for this vapor, however, the rate of vapor production is limited by the rate at which heat can be conducted to the bubble wall in order to satisfy the latent heat requirements. This situation is depicted in Figure 2b. The bubble wall velocity is positive (growth), the bubble wall temperature T_l and vapor pressure P_v have dropped from their initial values (cooling due to vaporization) and thus a driving force $T_\infty - T_l$ for heat transport has developed. The growth rate is determined by the pressure differential $P_l - P_\infty$, which is in turn determined by the opposing effects of bubble expansion and heat transport as previously discussed.

In real situations, the detailed manifestation of this physical phenomena can be quite complex. For example, the liquid phase temperatures may exhibit great temporal and spatial variations so that various portions of the bubble wall experience different temperature driving forces, possibly changing with time. An important example of this behavior is subcooled nucleate boiling, where as the bubble grows from the superheated wall region, it encounters subcooled liquid masses upon which condensation can take place. Thus portions of bubble wall could be evaporating while other portions could be condensing. This behavior, taking place at the microscopic level, could be important even for the nucleation process itself. Other difficulties exist. Bubbles growing rapidly on solid walls do not have time to rise, hence they obtain a hemispherical shape. Because of the non-slip condition on the wall as the bubble grows, a thin liquid film known as the microlayer is left behind (see Figure 3). This film can vaporize through its own superheat and by additional heating from the wall. Buoyancy forces also can play a role in promoting rise which affects bubble shape and mixing, complicating even further heat transport considerations. A large amount of literature is available for these and other problems.

A first step towards understanding the key physical interactions discussed in the two previous paragraph is to

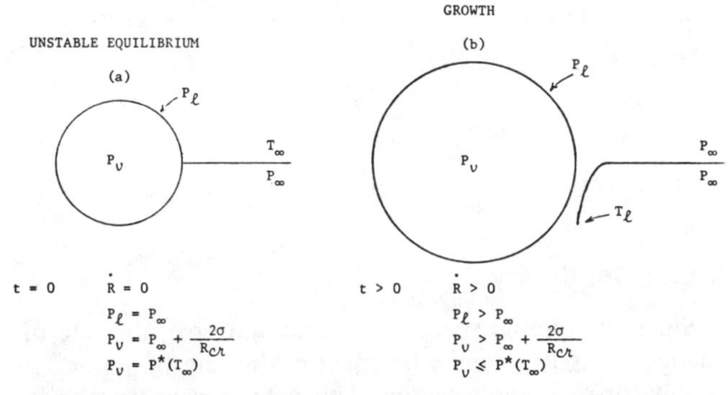

Figure 2. Qualitative illustration of vapor bubble growth from a cavity of critical size.

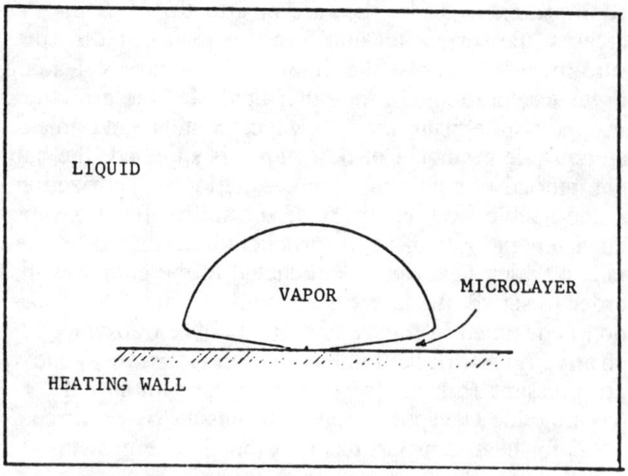

Figure 3. Illustration of microlayer geometry.

consider the problem of growth of a simple spherical vapor cavity, from its critical size, in a uniformly superheated incompressible liquid. Let us make the following not-so-restrictive assumptions:

a) The bubble remains stationary with respect to the liquid, which is of infinite extent. In fact, because of the spherical divergence, the liquid velocities decrease rapidly away from the bubble so that any walls away by a few bubble diameters are not felt, i.e., a finite system behaves as if it was infinite.
b) The bubble remains spherical throughout the growth period. This is basically a consequence of the previous assumption.
c) The vapor in the bubble remains saturated at all times and it is in equilibrium with the bubble wall. In most cases of interest this is the case.
d) Liquid viscosity is negligible. This is a good assumption except for some very high viscosity liquids.
e) The vapor-to-liquid density ratio is much smaller than unity. This is true except near the thermodynamic critical point.

With these assumptions, let us proceed to the mathematical formulation of the growth problem (refer to Figure 4 for notation).

Liquid Continuity

$$\frac{\partial}{\partial r}(r^2 u) = 0 \tag{5}$$

which integrates to:

$$r^2 u = R^2 \dot{R} \tag{6}$$

Vapor Continuity

Since the bubble space is taken as uniform, the rate of change of its mass must be equal to the rate of vapor supply through vaporization. This rate of vaporization is obtained by converting the sensible heat supply to heat of

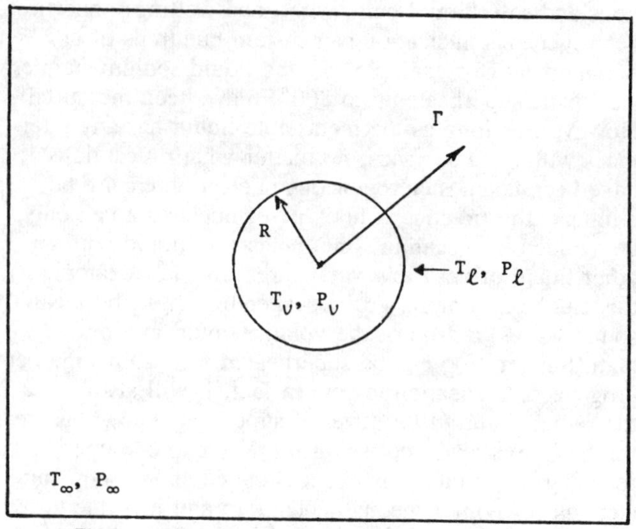

Figure 4. Notation for bubble growth formulation.

vaporization.

$$k_l \left.\frac{\partial T}{\partial r}\right|_R \frac{1}{\lambda} \tag{7}$$

where λ is the latent heat per unit mass. Hence:

$$\frac{d}{dt}\left(\frac{4}{3}\pi\rho_v R^3\right) = 4\pi R^2 \left\{ k \left.\frac{\partial T}{\partial r}\right|_R \frac{1}{\lambda} \right\} \tag{8}$$

Liquid Momentum

$$\frac{\partial u}{\partial t} + u\frac{\partial u}{\partial r} = -\frac{1}{\rho_l}\frac{\partial p}{\partial r}$$

at $t > 0$

$$R < r < \infty \tag{9}$$

with

$$P(R, t) = P_l$$
$$P(\infty, t) = P_\infty$$
$$u(r, 0) = 0$$
$$R(0) = R_{cr}$$

The radial velocity at infinity is zero (see Equation 6), hence it is convenient to integrate this equation over the radial coordinate to obtain an equation for the bubble wall. First use Equation 6:

$$\frac{1}{r^2}\frac{d}{dt}(R^2\dot{R}) + \frac{1}{2}\frac{\partial u^2}{\partial r} = -\frac{1}{\rho_l}\frac{\partial P}{\partial r} \tag{10}$$

Now integrate from R to ∞.

$$\frac{1}{R}\frac{d}{dt}(R^2\dot{R}) - \frac{1}{2}\dot{R}^2 = \frac{1}{\rho_l}(P_l - P_\infty) \tag{11}$$

Relate P_l to P_v by Equation 2 and rearrange:

$$R\ddot{R} + \frac{3}{2}\dot{R}^2 = \frac{1}{\rho_l}(P_v - P_\infty) - \frac{2\sigma}{\rho_l R} \tag{12}$$

at $t > 0$ with

$$\dot{R}(0)=0$$

$$R(0)=R_{cr}$$

Vapor Momentum

The need for this equation has been eliminated by assuming uniform vapor.

Liquid Energy

$$\frac{\partial T}{\partial t}+u\frac{\partial T}{\partial r}=\alpha_l\frac{1}{r^2}\frac{\partial}{\partial r}\left\{r^2\frac{\partial T}{\partial r}\right\}$$

$$t>0$$

$$R<r<\infty \tag{13}$$

or making use of Equation 6,

$$\frac{\partial T}{\partial r}+\frac{R^2\dot{R}}{r^2}\frac{\partial T}{\partial r}=\alpha_l\frac{1}{r^2}\frac{\partial}{\partial r}\left\{r^2\frac{\partial T}{\partial r}\right\} \tag{14}$$

with

$$T(r,\ 0)=T_\infty$$

$$T(R,\ t)=T^*(P_v)$$

$$T(\infty,\ t)=T_\infty$$

the second condition expresses assumption (c).

Vapor Energy

The need for this equation has been eliminated by assuming uniform, saturated vapor in equilibrium with the bubble wall.

System Coupling Discussion

Equations 12 and 14 are coupled through Equation 8. For a given P_v, Equations 12 and 14 can be solved for a small time increment for the acceleration (velocity and radius too) and temperature distribution in the liquid, respectively. These solutions can then be used in Equation 8 to find the resulting vapor density and thus, new vapor pressure (saturated vapor assumption). More elaborate models are possible, avoiding the assumptions of saturated vapor in equilibrium with the bubble wall, by making use of the vapor energy equation and a rate of vaporization expression based on the kinetic theory of gases; however, for most cases of interest these solutions indicate that this assumption is very nearly correct.

SOLUTIONS

Before considering the solution of the complete system of differential equations, it is instructive to consider certain limiting cases. These cases may be physically motivated and correspond to assigning total control to only one of the participating physical processes. For the case at hand, for example, one limit would correspond to negligible heat transfer limitations. The temperature of the bubble wall would then be essentially equal to T_∞ and thus, the pressure of the vapor would remain constant and equal to its initial value $P^*(T_\infty)$. In this limit, Equation 12 is sufficient to solve for the growth rate. Since momentum effects play the primary role in this limit, this solution will be called the "momentum solution." The other extreme would correspond to the case where heat transfer is so limiting that the bubble wall cools quickly to the lowest possible temperature. Since in the complete process the pressure of the vapor starts at its maximum value, $P^*(T_\infty)$ and monotonically decreases towards P_∞, the lowest possible vapor pressure is P_∞ and hence, the lowest possible wall temperature is $T^*(P_\infty)$. In this limit, the second boundary condition for Equation 14 is completely derived and thus a solution to Equation 14 can be obtained. Using this solution in Equation 8 the growth rate can be found. The solution at this limit will be called the "energy solution." The complete solution, discussed after the "momentum" and "energy" solutions, must encompass these two asymptotic limits and will provide quantitative criteria specifying the limits of their applicability.

Momentum Solution

The surface tension forces are important for the very initial (microscopic) stages of the growth process. The bubble growth process may be thought of to initiate from a small perturbation in any of the three quantities of the right-hand side of Equation 12, such as to produce a small positive result (initially the right-hand side is zero). According to Equation 12, an acceleration of the wall results, and this growth leads to a rapid decrease of the surface tension forces. Typically, critical radii are smaller than 10^{-3} cm, thus by the time the bubble is 10^{-1} cm the surface tension force has dropped to 1% of its initial value, i.e., can be considered negligible. Further, assuming that $P_v = P^*(T_\infty)$, and letting $\Delta P = P_v - P_\infty$, Equation 12 can be rearranged:

$$\frac{1}{2R^2\dot{R}}\frac{d}{dt}\{R^3\dot{R}^2\}=\frac{\Delta P}{\rho_l} \tag{15}$$

or

$$d\{R^3\dot{R}^2\}=2\frac{\Delta P}{\rho_l}R^2dR \tag{16}$$

which can be integrated to give:

$$R^3\dot{R}^2=\frac{2}{3}\frac{\Delta P}{\rho_l}\{R^3-R_{cr}^3\} \tag{17}$$

$$\therefore\ \dot{R}^2=\frac{2}{3}\frac{\Delta P}{\rho_l}\left\{1-\left(\frac{R_{cr}}{R}\right)^3\right\} \tag{18}$$

This equation indicates that the velocity of the wall increases rapidly from zero to a constant limiting value of

$$\dot{R}=\left\{\frac{2}{3}\frac{\Delta P}{\rho_l}\right\}^{1/2} \tag{19}$$

which is essentially obtained by the time the bubble has grown to five times the critical size. This means that the wall acceleration may be neglected in Equation 12, and Equation 19 is applicable for the whole growth process. From Equation 19, the bubble radius (take $R_{cr} \sim 0$) may be obtained as:

$$R = \left\{\frac{2}{3}\frac{\Delta P}{\rho_l}\right\}^{1/2} t \qquad (20)$$

Energy Solution

Because of the nonlinearity present in the convective term of Equation 14, an analytical solution is rather involved. Let us proceed, therefore, with the help of a heuristic argument. The problem does have the character of transient conduction in an infinite media, hence expect that the thermal boundary layer will vary proportional to $\sqrt{\alpha_l t}$, and the approximate temperature gradient at the bubble wall will be:

$$\left.\frac{\partial T}{\partial r}\right|_R = \frac{T_\infty - T^*(P_\infty)}{\sqrt{\alpha_l t}} \qquad (21)$$

Expanding the left-hand side of Equation 8 [using $\rho_v = \rho^*(P_\infty)$, which is constant] and utilizing Equation 21 gives:

$$4\pi\rho_v^*(P_\infty)R^2\dot{R} = 4\pi R^2 k_l \frac{T_\infty - T^*(P_\infty)}{\sqrt{\alpha_l t}}\frac{1}{\lambda} \qquad (22)$$

or

$$\dot{R} = \frac{k_l(T_\infty - T^*(P_\infty))}{\lambda\rho_v^*(P_\infty)\sqrt{\alpha_l t}} \qquad (23)$$

This equation in fact is within a constant of 0.98 from the exact solution of Equation 14. The solution for the radius (take $R_{cr} \sim 0$) is obtained by integration:

$$R = 2\frac{k_l[T_\infty - T^*(P_\infty)]}{\lambda\rho_v^*(P_\infty)\sqrt{\alpha_l}}\sqrt{t}$$

$$= \frac{2\rho_l\hat{c}_l[T_\infty - T^*(P_\infty)]}{\lambda\rho_v^*(P_\infty)}\sqrt{\alpha_l t} \qquad (24)$$

Coupled Solution

Assume that Equations 19 and 23 are valid over the regime of combined momentum and energy control, provided that the "actual" driving forces are utilized:

$$\frac{dR}{dt} = \left\{\frac{2}{3}\frac{P_v(t) - P_\infty}{P_l}\right\}^{1/2} \qquad (25)$$

$$\frac{dR}{dt} = \sqrt{\alpha_l}\frac{\rho_l c_l[T_\infty - T_l(t)]}{\lambda\rho_v^*(T_v)}\frac{1}{\sqrt{t}} \qquad (26)$$

Define

$$\phi^2 = \frac{T_v - T^*(P_\infty)}{T_\infty - T^*(P_\infty)} \qquad (27)$$

and relate $P_v(t)$ and ρ_v to T_v by means of linear interpolation between the initial state T_∞, $\rho_v^*(T_\infty)$ and the final state, $T^*(P_\infty)$, P_∞, that is

$$P_v - P_\infty = P_v^*(T_v) - P_\infty = [P_v^*(T_\infty) - P_\infty]\phi^2 \qquad (28)$$

$$\rho_v - \rho_v^*(P_\infty) = \rho_v^*(T_v) - \rho_v^*(P_\infty) = [\rho_v^*(T_\infty) - \rho_v^*(P_\infty)]\phi^2 \qquad (29)$$

Thus Equation 25 and 26 become:

$$\frac{dR}{dt} = A_*\phi \qquad (30)$$

$$\frac{dR}{dt} = \frac{B}{2}\frac{(1-\phi^2)}{[1+(\Gamma-1)\phi^2]}\frac{1}{\sqrt{t}} \qquad (31)$$

where

$$A_*^2 = \frac{2}{3}\frac{P^*(T_\infty) - P_\infty}{\rho_l} \qquad (32)$$

$$B^2 = 4\alpha_l\left\{\frac{\rho_l c_l[T_\infty - T^*(P_\infty)]}{\lambda\rho_v^*(P_\infty)}\right\}^2 \qquad (33)$$

and

$$\Gamma = \frac{\rho_v^*(T_\infty)}{\rho_v^*(P_\infty)} \qquad (34)$$

The group A_* has dimensions length over time and the group B^2 has dimensions length square over time. They can be utilized, therefore, to define characteristic length and velocity scales, and thus nondimensionalize the system (Equations 30 and 31).

$$R^* = \frac{R}{(B^2/A_*)}$$

and

$$t^* = \frac{t}{(B^2/A_*^2)} \qquad (35)$$

$$\therefore \frac{dR^*}{dt^*} = \phi \qquad (36)$$

$$\frac{dR^*}{dt^*} = \frac{1-\phi^2}{2[1+(\Gamma-1)\phi^2]}\frac{1}{\sqrt{t^*}} \qquad (37)$$

From Equations 36 and 37, for ϕ:

$$\frac{\phi}{(1-\phi^2)}[1+(\Gamma-1)\phi^2] = \frac{1}{2\sqrt{t^*}} \qquad (38)$$

which can be solved for $\phi = \phi(t^*; \Gamma)$. Equation 36 can then be integrated to yield R^* as a function of t^* with Γ as a parameter, i.e., $R^* = R^*(t^*; \Gamma)$. The results are plotted in Figures 5 and 6. From these solutions, for $t^* < 10^{-2}$, a good approximation is $R^* = t^*$, which is the same as the momentum solution Equation 20; for $t^* >$

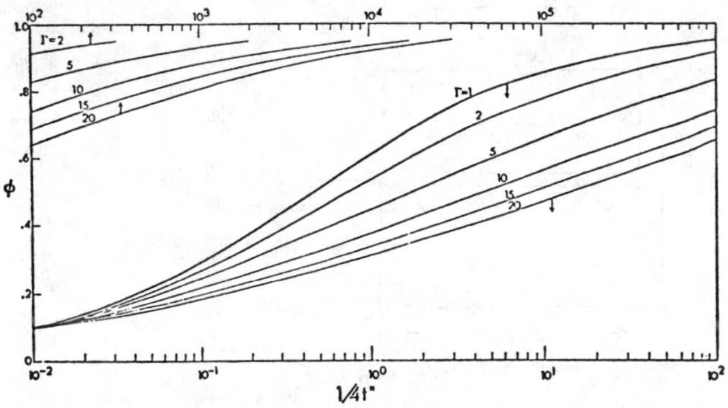

Figure 5. Solution to Equation 38.

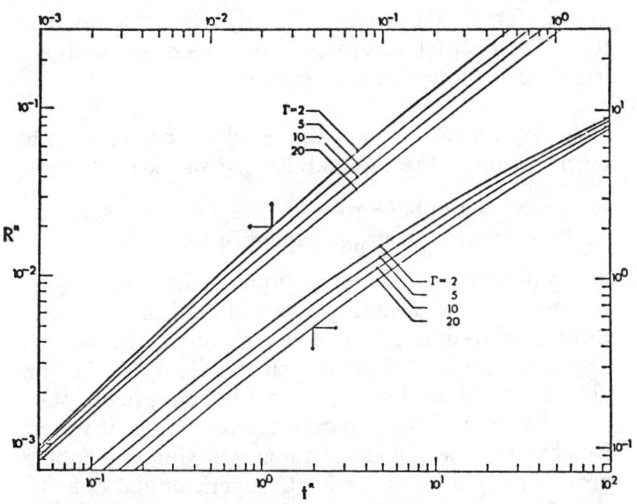

Figure 6. Universal plot of the bubble radius versus time.

10^2, a good approximation is $R^* = \sqrt{t^*}$, which is the same as the energy solution Equation 24. For intermediate values of t^*, the process is controlled by both momentum and energy effects and the solution can be read off the graphs. This solution has been checked against experimental data as well as elaborate numerical simulations with excellent results.

For low values of the density ratio, ($\Gamma < 2$), the solution may be expressed analytically by:

$$R^* = \frac{2}{3B}\{(B^2t^*+1)^{3/2} - B^2t^*)^{3/2} - 1\} \tag{39}$$

where

$$B^2 = \frac{\lambda\rho_v^*(P_\infty)}{T^*(P_\infty)}\,\frac{T_\infty - T^*(P_\infty)}{P^*(T_\infty) - P_\infty} \tag{40}$$

NOMENCLATURE

A	=	frequency factor, 10^{32} cm^{-3} s^{-1}
A_*	=	see Equation 32
B	=	see Equation 33
$\hat{c}$	=	heat capacity
J	=	bubble nucleation rate
K	=	Boltzman's constant
k	=	thermal conductivity
P	=	pressure
ΔP	=	pressure difference
$P^*(T)$	=	saturation pressure at T
R	=	bubble radius
R^*	=	dimensionless radius, see Equation 35
$\dot{R} = dR/dt$	=	bubble wall velocity
$\ddot{R} = d^2t/dt^2$	=	bubble wall acceleration
r	=	radial coordinate
T	=	temperature
$T^*(P)$	=	saturation temperature at P
T_{es}	=	spontaneous nucleation temperature
t	=	time
t^*	=	dimensionless time, see Equation 35
u	=	radial liquid velocity

Greek Letters

α	=	thermal diffusivity
λ	=	latent heat of vaporization
Γ	=	see Equation 34
ρ	=	density
$\rho_v^*(P)$	=	saturation vapor density at P
$\rho_v^*(T)$	=	saturation velocity density at T
σ	=	surface tension
ϕ^2	=	see Equation 27

Subscripts

cr	=	critical
i	=	bubble internal
l	=	liquid
m	=	maximum
v	=	vapor
∞	=	condition at infinity

SUGGESTED COMPLIMENTARY READING

1. Dwyer, O. E., "Boiling Liquid-Metal Heat Transfer," Chapters 1–4, American Nuclear Society, Hinsdale, IL (1976).
2. Hsu, Y. Y., "Transport Processes in Boiling and Two-Phase Systems," Chapters 1 and 2, McGraw Hill, New York (1976).

STUDY PROBLEMS

1. In the absence of dissipation, the PdV work done by the vapor during bubble growth shows up as kinetic energy of the liquid. Utilize this idea to obtain an equation for the bubble wall motion. Now discuss the possibility of growth of a cylindrical or plane bubble in an infinity sea of liquid.
2. Consider the effects of viscosity on bubble growth rates.
   (1) Should viscosity appear in the equation of motion, Equation 9? If so, in what form? If not, is the flow inviscid or irrotational?
   (2) Modify Equation 12 to take into account viscous effects, by utilizing an augmented form of Equation 2 that equates the total radial stress in the liquid (at the wall) to $P_v - 2\sigma/R$.
   (3) Consider the order of magnitude comparison of the various terms in the modified Equation 12, to obtain a criterion for judging the importance of viscous effects.

3. Obtain approximate "energy solutions" for the bubble growth problem by neglecting appropriate terms in the thermal energy equation. Compare and discuss in physical terms the observed differences.

4. Obtain an approximate "energy solution" for the bubble growth problem by means of the integral method. Utilize a quadratic temperature profile. Compare to the exact solution.

HOMEWORK PROBLEMS

1. Calculate the size of the largest cavity present for a water system (at atmospheric pressure) that nucleates 50°C below its spontaneous nucleation limit. Estimate the resulting bubble growth rate, in miles per hour. At what time will the growth become controlled by thermal effects? At what time will 50% of the maximum possible driving force for heat transfer develop?

2. A glass apparatus is being constructed for the study of nucleation and macroscopic growth of vapor bubbles in highly superheated liquids (Figure 7). You are asked to review the design and determine whether any safety problems exist. You'd also want to examine whether any operating limits can be specified to assure safety.

   Intended Design and Operation: The apparatus consists of a cylindrical glass vessel as shown in the figure.

   The nucleation point is approximately at the center of the flask. The flange A is used for the various penetrations such as vacuum give, pressure tap, etc. (not shown). The glycerol is used to prevent premature nucleation. Infrared heating can be used to bring *quickly* and uniformly the water up to any desired temperature. The desired superheat is then recorded with a high speed motion picture camera.

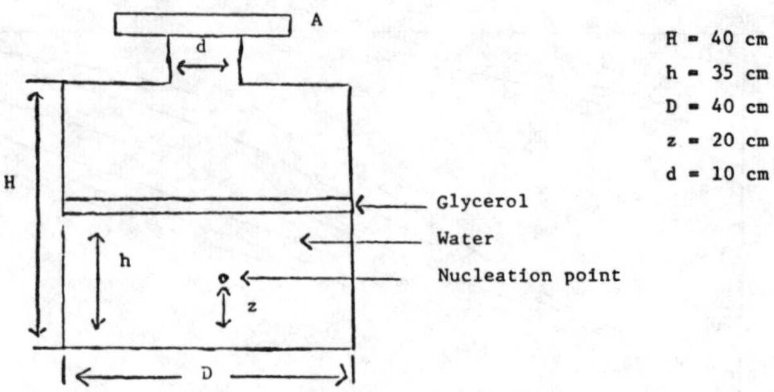

Figure 7.

   Property specs: The apparatus (glass) can take up to 60 psi primary loading and 200 psi shock loading.

   Assume: The bubble grows in a spherically symmetric fashion with no displacement of its center.

3. An important application of liquid sodium is as a coolant in some advanced types of nuclear reactors. Boiling of sodium plays an important role in the safety assessment of these systems. Consider the simulation of sodium boiling from a bubble growth standpoint by the much more convenient boiling of water. Determine water conditions that will simulate bubble growth in sodium under 30°C superheat and one atmosphere pressure. Discuss the results in physical terms.

Module C6.2

Miscible Dispersion

R. Shankar Subramanian

Department of Chemical Engineering
Clarkson University
Potsdam, New York 13676

OBJECTIVES
After completion of this module, the student should be able to describe and discuss the effects of convective transport and diffusive transport on the axial dispersion of solute in a conduit.

PREREQUISITE MATHEMATICAL SKILLS
1. Elementary calculus and differential equations.
2. An introduction to partial differential equations, such as the diffusion equation. Previous exposure to the equation of conservation of mass would be helpful.

PREREQUISITE ENGINEERING AND SCIENCE SKILLS
1. Elementary modeling of transport problems, especially elementary concepts of convective and diffusive transport.

INTRODUCTION

Miscible dispersion refers to the mixing phenomena which occur when two miscible fluids are brought together either when stationary or when they are in motion. This definition is very broad, and it is not possible to do justice to the wide range of topics it covers in one module. So, attention will be confined here to the phenomenon of dispersion which takes place when a solute is introduced into fluid flowing in a conduit. Specifically, no attempt will be made to treat mixing in agitated vessels, which also would be called "dispersion."

While dispersion phenomena are fascinating to study just from the point of view of scientific interest, there are practical reasons for developing a good understanding of such phenomena. Some applications are:

1. The distribution of tracers and drugs in the bloodstream.
2. Conventional chromatography.
3. Field flow fractionation or polarization chromatography.
4. The transient behavior of tubular reactors.
5. Pollutant transport in the atmosphere.
6. Material and thermal pollution of natural streams.

The list is by no means complete. New applications will, doubtless, be found in the future. Serious study of the subject was begun only about thirty years ago by G. I. Taylor in a classic paper (*52*), and new developments are being made even today.

In each of these examples, it might be seen that dispersion or mixing occurs due to two basic mechanisms. These are:

1. convective or bulk transport
2. diffusive or molecular transport

Simplistic as this might sound, the interactions between these two mechanisms lead to some very interesting consequences. Before proceeding further, it is important to distinguish between two basically different types of dispersion problems which emerge from the various applications.

In the first category, called "initial distribution problems," a certain finite amount of solute is introduced into a flow and permitted to disperse. There can be no nontrivial steady state in such a situation, and all phenomena of interest are transient. Elution chromatography is a good example of this problem. In generalized dispersion theory, such problems are handled by using an infinite series solution which satisfies the appropriate initial and boundary conditions. This series will be displayed later in this module.

A second category of dispersion problems is characterized by the example of a chimney emitting pollutant continuously into the atmosphere or a chemical plant discharging effluent continuously into a nearby stream. In such problems, one might imagine the fluid medium (atmosphere or river) to be initially devoid of solute. At time zero, the appropriate source of solute is turned on at the inlet. This type of problem may be termed an "inlet distribution problem." If the discharge rate is steady, a non-trivial steady state can be established at locations close to the system inlet after a while. This steady state will propagate down the flow as time goes on, carrying in front of it a transient region. One approach for handling this class of problems is to treat the continuous source as a series of pulses, each one of which can be handled using techniques suitable for initial distribution problems. This method is known as superposition (*13*) and its application to inlet distribution problems has been discussed by Gill and Sankarasubramanian* (*25*, *41*), and Gill (*26*), and clarified by Subramanian (*51*). More recently, Smith (*43–46*) has used superposition innovatively to construct solutions to both classes of problems using his "delay-dispersion model."

In the rest of this module, attention is focused on "initial distribution problems" because of their relative simplicity.

THE PHYSICAL PROBLEM

The dispersion of a solute introduced into a fluid flowing into a conduit will be considered here. It is convenient to approach the complex problem in which a solute is dispersed due to the simultaneous action of convection and diffusion in the presence of a non-uniformly distributed velocity field, say in a circular tube, via simple, idealized, intermediate physical situations. These might be called "thought experiments."

Consider a slug of dilute solution, x_s units long and having a uniform concentration C_o of a suitable solute held in an infinitely long circular tube as shown in Figure 1. This slug is separated from pure solvent on either side of it by suitable impermeable barriers. The axial coordinate x and the radial coordinate r are defined as shown in the figure. All the thought experiments will be performed on this system.

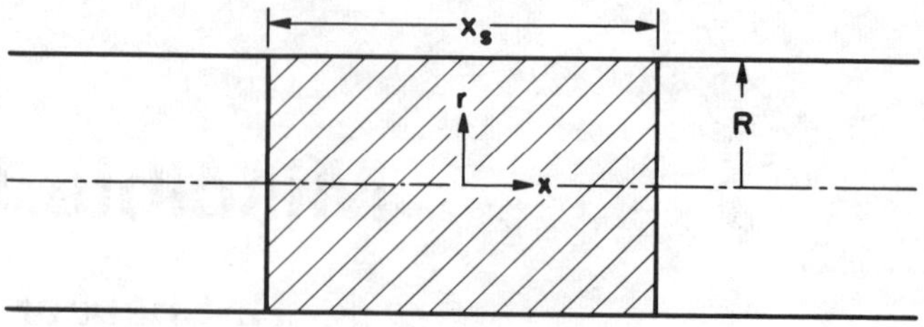

Figure 1. The system.

Experiment 1

At time zero, imagine the barriers on either side of the slug are removed without causing any disturbance in the fluid (this is very hard to do in practice, but easy to imagine). No pressure gradient is imposed on the fluid so that it remains completely stationary.

The solute will be observed to diffuse symmetrically on either side of the original slug. The velocity resulting from this diffusion is ignored by assuming the slug to be very dilute. There is no radial or angular variation of concentration in the system, since the slug is assumed to be initially uniform in concentration everywhere. This diffusion process can be described mathematically by the one-dimensional conservation equation for the local solute concentration, $C(t, x)$. For a derivation of the conservation equation, see Bird, et al (*6*).

$$\frac{\partial C}{\partial t} = D \frac{\partial^2 C}{\partial x^2} \tag{1}$$

Here, t is time, and D is the molecular diffusivity of the solute in the fluid. The initial and boundary conditions on $C(t, x)$ for the system are:

$$\left.\begin{aligned} C(0, x) &= C_o, \quad |x| \le \tfrac{1}{2} x_s \\ &= 0, \quad |x| > \tfrac{1}{2} x_s \end{aligned}\right\} \tag{2}$$

$$C(t, \pm\infty) = 0, \quad t \ge 0 \tag{3}$$

Equation 2 states that initially, the solute concentration is C_o everywhere inside the slug and zero everywhere else. Equation 3 is characteristic of problems of this type, and indicates that solute never reaches axial stations infinitely far away from $x = 0$. The solution of Equations 1 to 3 for $C(t, x)$ is well-known and is given in Crank (*14*). Figure 2 shows the concentration distribution from this solution as a function of axial position for various values of time. In this simple thought experiment, there is no bulk (convective) transport. Figure 2 shows that symmetric axial spreading or dispersion of solute occurs due to molecular diffusion.

Experiment 2

Just as was done in the first experiment, imagine the barriers on either side of the slug are removed at time zero without causing any disturbance. However, this time, an axial pressure gradient is imposed on the fluid. This will result in bulk motion. This motion is assumed to be described by an idealized plug flow velocity profile. That is,

$$v_x = \text{constant} = v_m \tag{4}$$

Here, v_m is the cross-sectional average velocity defined by

$$v_m = \frac{\int_0^R v_x r \, dr}{\int_0^R r \, dr} = \frac{2}{R^2} \int_0^R v_x r \, dr \tag{5}$$

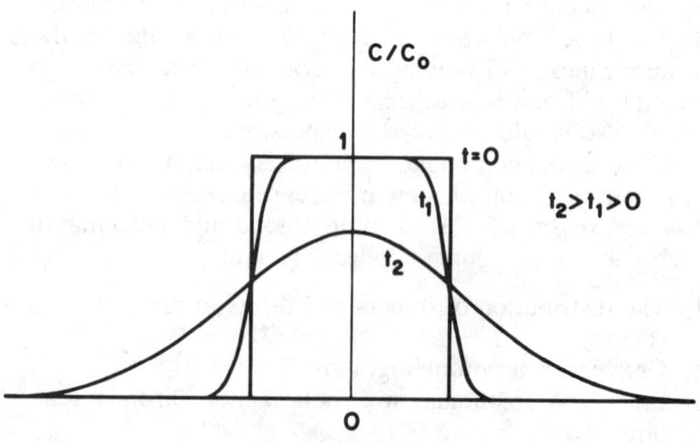

Figure 2. Concentration versus axial coordinate for the diffusion of a slug.

*R. Shankar Subramanian was formerly known as R. Sankarasubramanian.

Since v_x is independent of position in this thought experiment, $v_m = v_x$. The fluid, and hence the solute, is moving with a constant velocity v_m everywhere. So, an observer who starts at time zero at $x = 0$, and runs along with the slug (moves axially at the same velocity v_m), should be able to "freeze" the motion of the solute. Then, the only transport seen by this observer would be molecular diffusion. That is, in a coordinate system (t, x_1) where

$$x_1 = x - v_m t \tag{6}$$

$C(t, x_1)$ would satisfy a simple diffusion equation. For this thought experiment, the concentration profiles would be identical to those in Figure 2 if the abcissa is changed from x to x_1.

The lesson to be learned from this thought experiment is that a uniform velocity field has no effect on axial spreading. The only role such bulk motion plays is to physically translate the solute slug with a velocity v_m down the tube. The center of the solute distribution is always located at $x = v_m t$ or $x_1 = 0$. Solute spreads symmetrically about this center by molecular diffusion only. For completeness, the partial differential equation satisfied by the solute concentration in the (t, x_1) system is:

$$\frac{\partial C}{\partial t} = D \frac{\partial^2 C}{\partial x_1^2} \tag{7}$$

Note that Equation 7 is identical to the diffusion equation for C in the first thought experiment except for the fact that x_1 is measured from an origin moving at a speed v_m.

Experiment 3

This experiment is similar to Experiment 2, but the system is made more realistic by assuming laminar flow to exist instead of the idealized plug flow. That is, at time zero, the barriers are removed, and a pressure gradient is applied which causes steady, fully developed laminar flow in the tube. From the solution of the momentum conservation and total mass conservation equations for this system (*6*), it is known that the velocity is distributed non-uniformly across the cross-section of the tube in the shape of a parabola. That is,

$$v_x(r) = 2v_m\left(1 - \frac{r^2}{R^2}\right) \tag{8}$$

where r is the radial coordinate and R is the tube radius. v_m still represents the average velocity as can be verified from Equations 5 and 8.

From Equation 8, it may be observed that the velocity has a maximum value of $v_x = 2v_m$ at the tube center, and is zero at the tube wall. Clearly, it is impossible for an observer to "freeze" this motion by running along at any particular velocity as was done in the earlier thought experiment. The effect of the r-dependent velocity field is to cause *radial* concentration gradients in the tube even though the slug is initially radially uniform in concentration. Figure 3 shows the initial slug, and the condition of the slug after a small amount of time has elapsed.

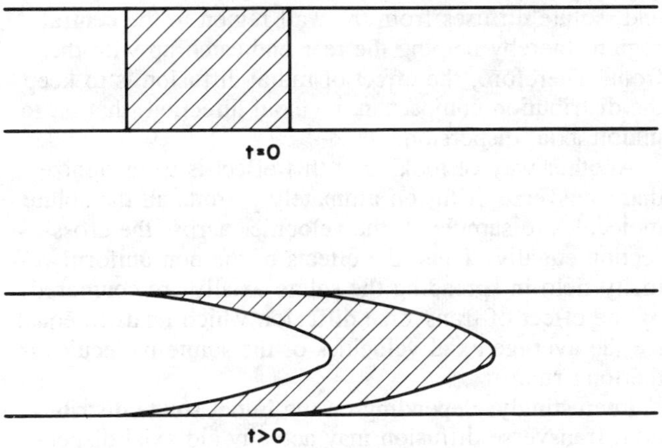

Figure 3. Axial spreading of the solute due to convection.

The front end of the slug starts out at time zero as a flat circular surface. However, due to the action of the parabolic velocity field, this surface becomes paraboloidal in shape. Similarly, the rear end of the slug becomes another paraboloidal surface. Ignoring diffusion for the time being, all the solute is located in the region between these two surfaces.

It might be noted immediately that the solute has been axially spread out just by the convection process. That is, the axial region occupied by the solute has been increased by the effect of the non-uniform velocity field. In fact, if there were no diffusional effects, this spreading process would go on distributing the solute over increasingly larger axial extents. Thus, axial dispersion is enhanced by the convective transport process in this example.

The role of diffusion in this problem is interesting and counterintuitive. One normally thinks of diffusion as a spreading mechanism. Yet, as will be seen shortly, the principal role of diffusion here is to inhibit the spreading induced by the convective transport process.

In Figure 3, the diffusive movement of solute will be normal to the surfaces which separate solute from the region devoid of solute. This diffusive flux may conveniently be broken up into an axial component (in the x-direction) and a transverse component (in the r-direction). The effects of these diffusive fluxes on the axial spreading of solute is opposite. Axial diffusion, as noted in the earlier examples, contributes to axial spreading. However, transverse diffusion inhibits axial spreading. Figure 4 illustrates why.

Shown in Figure 4 are the directions of radial solute transport due to diffusion. All over the front end of the distribution, radial diffusion moves solute from the faster central region to the slower region near the wall. Thus, the front end of the distribution cannot move ahead as rapidly as it would if there was no diffusion. At the rear

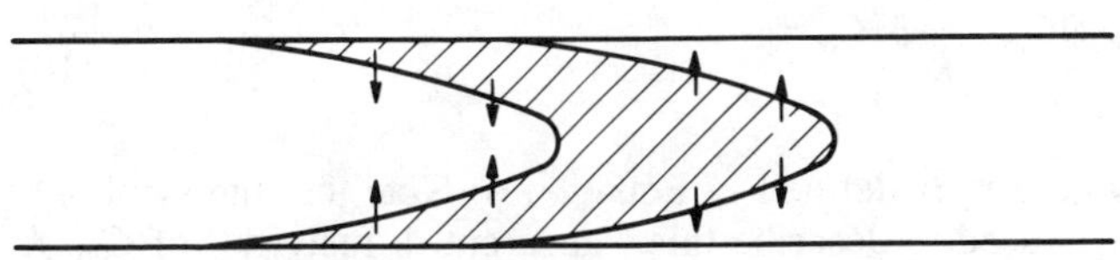

Figure 4. The role of transverse diffusion.

end, solute diffuses from the wall region to the central region, thereby helping the rear end catch up with the front. Therefore, the effect of radial diffusion is to keep the distribution compact in the axial direction, that is, to inhibit axial dispersion.

Another way of looking at this effect is to recognize that transverse diffusion ultimately permits all the solute molecules to sample all the velocities across the cross-section equally. Thus, the effects of the non-uniform velocity field in spreading the solute axially are countered by the effect of transverse diffusion which tends to equalize the average axial velocities of the solute molecules in the long run.

Interestingly, depending on the initial solute distribution, transverse diffusion may actually aid axial dispersion in the initial stages in certain situations (*24*). However, given time scales on the order of the diffusion time across the tube ($\sim R^2/D$), the effect of transverse diffusion is to inhibit axial spreading.

In summary, the third experiment performed under laminar flow conditions shows that convective transport in the form of a non-uniform velocity field tends to enhance axial dispersion; axial diffusion aids such axial dispersion, and transverse diffusion inhibits axial dispersion.

Unlike its predecessor, Experiment 3 can actually be set up in a laboratory. The results are fascinating when one does so. Near the beginning of this century, Griffiths (*27*) performed studies of the dispersion of a drop of fluorescent solution in a stream of water in laminar flow through a cylindrical capillary. *He found experimentally that the dye spreads out symmetrically about a plane which moves with the average flow velocity of the water in the tube.* This is a surprising observation since the velocity profile is asymmetric about the streamline in which the velocity is equal to the average velocity of flow. It was left to G. I. Taylor (*52*) to perform additional experiments and explain what Griffiths saw by means of a series of intuitive arguments. In view of the limited space available here, it is not possible to do justice to Taylor's work on this subject. However, the interested reader may consult his paper for such details. One should first recognize that when the dye concentration in a tube is visually examined, the observed average concentration is a cross-sectional average C_m. This is defined as:

$$C_m(t, x) = \frac{\int_0^R C(t, x, r) r\, dr}{\int_0^R r\, dr} = \frac{2}{R^2} \int_0^R Cr\, dr \tag{9}$$

Recalling the second thought experiment, if an observer runs along with the mean velocity of flow v_m in Griffiths' experiments, according to Griffiths, the observer would see what *appears* to be a diffusive process as far as the average concentration is concerned. That is,

$$\frac{\partial C_m}{\partial t} = k \frac{\partial^2 C_m}{\partial x_1^2} \tag{10}$$

where x_1 is defined in Equation 6. Note that the symbol k is used to describe this "apparent diffusivity" of C_m. k cannot be simply equal to the molecular diffusivity D since the axial spreading of solute in Griffiths' experiments is due to the complex interaction between convective and diffusive transport, and not due to simple molecular diffusion. In fact, it is worth noting that the "apparent diffusivity" is typically several orders of magnitude larger than molecular diffusivity.

Taylor (*52*) performed some experiments of his own, as mentioned earlier (using potassium permanganate as the solute and water as the solvent) and confirmed Griffiths' observations. He also developed an approximate theory of the process (ignoring axial diffusion) which led to Equation 10 with the following result for the "dispersion coefficient" or the "apparent diffusivity," k.

$$k = \frac{R^2 v_m^2}{48D} \tag{11}$$

Taylor's theoretical result is in agreement with the qualitative arguments presented in the earlier discussion of Experiment 3. The dispersion coefficient, k, which represents the axial spreading of solute in the capillary, is directly proportional to the square of the average velocity. The larger the average velocity, the greater will be the absolute velocity variations across the capillary, and therefore, the greater will be the axial spreading due to the velocity profile effect (convective spreading). Furthermore, the dispersion coefficient is inversely proportional to the molecular diffusivity D. This also is in agreement with the reasoning presented earlier which indicated that transverse diffusion should inhibit axial spreading.

Taylor's description of the dispersion process is valid only after a sufficient amount of time has elapsed since the introduction of solute into the capillary. This time scale is on the order of the diffusion time across the capillary R^2/D. Of course, several interesting transient phenomena take place before this time scale is reached in reacting systems. Also, in short conduits, the solute will leave the conduit before this time scale is reached. A generalized theory which accounts for solute behavior from time zero was developed by Gill and Sankarasubramanian (*23, 24*), who showed that the result for the dispersion coefficient derived by Taylor and displayed above is asymptotically valid for times $t \gtrsim 0.5\, R^2/D$.

Taylor's result, given in Equation 11, is based on the assumption that axial molecular diffusion can be neglected. Aris (*1*) using a different technique (the method of moments), showed that the effect of axial molecular diffusion on the Taylor dispersion coefficient is additive. That is,

$$k = D + \frac{R^2 v_m^2}{48D} \tag{12}$$

when one includes axial diffusion in the theory. Interestingly, the theory of Aris is indeed valid for all values of time since the introduction of the solute. However, emphasis in his work was directed at obtaining large time asymptotes consistent with Taylor's observations.

Equation 12 is known widely as the Taylor-Aris result for the dispersion coefficient in a tube.

ELEMENTS OF DISPERSION THEORY

The transport of solute in Griffiths's and Taylor's experiments may be described under suitable assumptions by the following mass conservation equation.

$$\frac{\partial C}{\partial t}+v_x(r)\frac{\partial C}{\partial x}=D\left[\frac{1}{r}\frac{\partial}{\partial r}r\frac{\partial C}{\partial r}+\frac{\partial^2 C}{\partial x^2}\right] \tag{13}$$

For an initial slug of solute of uniform concentration C_o, the following initial and boundary conditions may be written for the solute concentration $C(t, x, r)$.

$$\left.\begin{aligned} C(0, x, r) &= C_o, \quad |x| \leq \frac{1}{2}x_s \\ &= 0, \quad |x| > \frac{1}{2}x_s \end{aligned}\right\} \tag{14}$$

$$C(t, \pm\infty, r)=0 \tag{15}$$

$$C(t, x, 0)=\text{finite} \tag{16}$$

$$\left.\frac{\partial C}{\partial r}\right|_{r=R}=0 \tag{17}$$

Equations 14 and 15 are familiar ones. Equation 16 indicates that the concentrations in the system are finite, and Equation 17 results from the fact that the tube wall is impermeable to the solute. Even though Equations 13 through 17 present a formidable problem to tackle analytically, several methods of solution are available. One may use any one of these methods to compute $C(t, x, r)$, and from this, any desired average, such as the cross-sectional average $C_m(t, x)$ or the cup-mixing or bulk average $C_b(t, x)$. However, a certain class of solution techniques permits the user to derive equations for the average concentration distribution with coefficients which may be calculated from first principles without ever computing the local concentration field. In the rest of this section, an elementary version of this method is presented, and some generalizations are pointed out. For an introduction to other approaches, the reader may consult some of the literature cited in the bibliography. Examples include the papers of Barton (*2–4*) Lighthill (*33*), Chatwin (*7–12*), Tseng and Besant (*55, 56*), Fife and Nicholes (*21*), and Yu (*58*).

Taylor's work (*52*) is an example where Equation 10 for $C_m(t, x)$ and Equation 11 for k permit the user to *predict* the cross-sectional average concentration field from theory without ever calculating $C(t, x, r)$. His approach was based on intuition. He recognized that for large values of time, the difference between the local concentration C and the cross-sectional average concentration C_m must be small. He also recognized that these small radial variations in C are crucial in accounting for axial dispersion, and attempted to describe them by writing, in effect,

$$C-C_m \approx g(r)\frac{\partial C_m}{\partial x} \tag{18}$$

Using Equation 18, one may determine $g(r)$ and then go on to derive the dispersion model, Equation 10 with k given by Equation 11. Equation 18 is based on the assumption that the higher axial derivatives of C_m are smaller than the first.

Later, in (*54*), Taylor added a correction term.

$$C \approx C_m+g_1(r)\frac{\partial C_m}{\partial x}+g_2(r)\frac{\partial^2 C_m}{\partial x^2} \tag{19}$$

This has subsequently been developed into an infinite series by Gill and Sankarasubramanian (*22–24*) in "generalized dispersion theory." (See Appendix.)

$$C=\sum_{k=0}^{\infty} g_k(t, r)\frac{\partial^k C_m}{\partial x^k} \tag{20}$$

An important difference between Equations 19 and 20, in addition to the infinite series generalization, is the introduction of time dependence in the functions g_k. Taylor's theory applied only for large values of time because the initial condition cannot be satisfied even with an infinite series generalization of Equation 19. As pointed out earlier, in several problems, the phenomena of interest occur before such values of time are reached. The functions $g_k(t, r)$ are time dependent precisely for accounting for such phenomena.

Another interesting feature to note is that in Equation 20, g_o can in general depend on time and radial position. In contrast, $g_o = 1$ in Taylor's work. The more general g_o is important in several practical problems where the initial solute distribution is non-uniform in r [as when a syringe and needle are used to inject a slug into a larger diameter capillary (*24*)]. Also, in cases where there is mass transfer at the tube wall, say due to a chemical reaction, g_o reaches steady but radially non-uniform asymptotic representations at large values of time (*42*). This is also the case when there is transverse flow, as shown in Doshi, et al (*19*), or when the solute migrates in a transverse field (*31, 40*).

For actual details of the development of generalized dispersion theory and its applications, the reader may refer to the series of papers by Gill and Subramanian listed in the bibliography. It may be mentioned here that, in general, Equation 20 leads to the following result:

$$\frac{\partial C_m}{\partial t}=\sum_{i=0}^{\infty} k_i(t)\frac{\partial^i C_m}{\partial x^i} \tag{21}$$

Equation 21 is a generalization of Taylor's result given in Equation 10, and is, in fact, a consequence of using Equation 20 for C instead of the approximation implied in Equation 18. The coefficients $k_i(t)$ in the generalized dispersion equation can be calculated from first principles fairly easily when $k_o = 0$, which is the case when the solute flux is zero at the tube wall. More complex cases involving reactions at the tube wall, and situations where other transverse average concentrations (such as the bulk-average) are required can be handled by the method of DeGance and Johns (*15–17*).

It is important to provide some physical insight into

the nature of the coefficients appearing in Equation 21. k_o may be thought of as a pseudo-homogeneous reaction rate constant accounting for disappearance of solute through possible reactions at the wall. Or as in Doshi, et al (*19*) where loss of fluid through the walls results in a concentration buildup within the channel, k_o accounts for this process. The quantity $-k_1$ may be thought of as the velocity of the solute cloud through the tube. It is time-dependent in general. For instance, if the solute is injected only into the central regions of the tube in Experiment 3, initially the solute cloud will move rapidly. However, as the solute diffuses across the cross-section, asymptotically, in that experiment, the solute will move with the average speed of the flow. Thus, as $t \to \infty$, $k_1 \to -v_m$.

The dispersion coefficient, $k_2(t)$, in Equation 21, is inherently time-dependent. It starts out at time zero being exactly equal to the molecular diffusivity, D. k_2 increases with time, and asymptotically approaches the Taylor-Aris value given in Equation 12 (*23, 24*). In a short tube, this asymptotic condition will never be reached, and the small-time behavior of the dispersion coefficient can be very important. A case in point is the study of Ultman and Blatman (*57*), who examined the dispersion of solutes in models of the human lung. Here, relatively short segments of tube bifurcate into new tubes, and the phenomena of interest in each tube occur over a time scale that is short compared to the diffusion time scale across the tube.

The higher–order coefficients $k_i(t)$ for $i \geq 3$ in Equation 21 are small, and the result is usually truncated after the term involving k_2, to give the classical Taylor dispersion equation. However, at small values of time, the higher–order terms can be important.

An interesting new development in dispersion theory has been presented over the last few years by R. Smith in a series of papers (*43–46*). He introduces a solution form, different from Equation 20, which is in a natural way able to accommodate continuous discharges of solute, as well as instantaneous ones. The approach, described by Smith as a delay-diffusion model, holds much promise. In particular, it is known from numerical calculations that the axial distributions of the cross-sectional average concentration are actually skewed in the initial stages of thought Experiment 3. Thus, a dispersion model of the type in Equation 10 cannot predict such distributions due to its inherent spatial symmetry. Smith's delay-diffusion model does very well in predicting such distributions properly.

The ideas presented in conjunction with the generalized dispersion Equation 21 can be connected logically to the development of Aris (*1*), who derived results for the spatial moments of the average solute concentration. In particular, the time dependences of the speed of the solute cloud and the dispersion coefficient are anticipated by Aris in the results displayed by him for the moments. The connection among the coefficients $k_i(t)$ and axial moments of the solute distribution is given by Subramanian (*47*).

The nature and properties of the series in the right-hand side of Equation 20 are not obvious. While it is clearly a generalization of Taylor's result in Equation 19, it is not at all clear whether the series actually represents the left-hand side at least in some asymptotic sense. The question of the accuracy of truncated versions of Equation 20 among other things has been addressed by DeGance and Johns (*15*). The origin of Equation 20 becomes more clear if one examines the problem after taking Fourier transforms in the axial coordinate. This point is addressed briefly in the appendix.

As seen here, dispersion theory involves averaging the concentration over the cross-section and examining the behavior of this average as a function of the remaining variables. One might think of this as a general methodology wherein a function is averaged over "local" variables and the behavior of this average as a function of the remaining "global" variables is studied. This general point of view is put forth by Brenner (*5*), and is applicable to a wide variety of problems which superficially appear very different from the ones treated here. An example considered by Brenner in (*5*), as well as in several other papers which he has published recently, is that of the settling of a collection of non-spherical particles. Here, averaging is done over the various possible orientations the particles can assume, just as the concentration is averaged over the cross-section in the problem considered in this module.

Finally, a note of caution. The reader who goes to Taylor's classic paper (*52*) will find that Taylor used an intuitive procedure, not always rigorous, in arriving at the result for the dispersion coefficient k. He was successful in capturing the essence of this problem, as he had a knack of doing, because of his phenomenal insight. In particular, the axial derivatives of the average concentration are asymptotically small at large values of time. This permitted Taylor to make certain approximations, and correctly estimate the radial concentration variations which are crucial to the calculation of the asymptotic dispersion coefficient. Imitation of Taylor's procedure in a new setting without regard to rigor can easily lead the user to incorrect and indefensible predictions, a trap one must avoid.

CONCLUDING REMARKS

As demonstrated in this module, a simple experiment in the laboratory on the unsteady transport of a solute through a capillary can lead to interesting models. There are several areas which were left unaddressed in this module. The reader is encouraged to go directly to the published literature for more information. An excellent review article on the subject of dispersion authored by Nunge and Gill (*34*) appeared in 1969. This reference presents a state-of-the-art review of the area at that time and covers subjects such as dispersion in turbulent flow and in various flow geometries, and the effects of buoyancy on dispersion.

The bibliography provided at the end of this module not only lists the articles cited here, but also includes several pertinent articles which have appeared in the literature since the appearance of the Nunge-Gill review. While it is by no means an exhaustive list, it should direct the reader to the relevant literature for further enlightenment on this subject.

APPENDIX

The purpose of this Appendix is to briefly address the question of the origin of the generalized dispersion solution given in Equation 20 of the module. The method of Fourier transforms is used for this purpose. This material is relegated to an Appendix since it is not necessary to be familiar with it in order to understand the contents of the module; yet, it may help in appreciating the material a little better.

The Fourier transform of $f(x)$ may be defined as follows (*28*).

$$\bar{f}(\beta)=\int_{-\infty}^{\infty} f(x)e^{-i\beta x}\, dx \tag{A-1}$$

Then, the inverse is given by

$$f(x)=\frac{1}{2\pi}\int_{-\infty}^{\infty} \bar{f}(\beta)e^{i\beta x}\, d\beta \tag{A-2}$$

By taking Fourier transforms of both sides of Equation 20 in the module and rearranging the result, one gets

$$\frac{\bar{C}(t,\ r;\ \beta)}{\bar{C}_m(t;\ \beta)}\sim\sum_{k=0}^{\infty} g_k(t,\ r)\ (i\beta)^k* \tag{A-3}$$

Thus, Equation 20 results from an attempt to expand the ratio of the Fourier transforms of the local and average concentrations in a power series in β. The properties of such series can be investigated conveniently by transforming the equations for $C(t, x, r)$ to Fourier space.

ACKNOWLEDGMENT

I am grateful to Professor William N. Gill for providing critical comments on an earlier version of this module.

NOMENCLATURE

C = concentration of solute
C_m = cross-sectional average concentration (see Equation 9)
C_o = initial concentration in solute slug
D = molecular diffusivity
g_k = functions appearing in Equation 20
k = dispersion coefficient
k_i = coefficients appearing in Equation 21
r = radial coordinate
R = tube radius
t = time
v_x = axial velocity of fluid
v_m = average axial velocity
x = axial coordinate
x_1 = transformed coordinate (see Equation 6)
x_s = length of solute slug

*The symbol ~ means that the right-hand side represents left-hand side in the sense of an asymptotic expansion.

LITERATURE CITED

1. Aris, R., *Proc. Roy. Soc. Lond. A235,* 67 (1956).
2. Barton, N. G., *J. Fluid Mech. 74,* 81 (1976); *74,* 91 (1976).
3. Barton, N. G., *J. Fluid Mech. 136,* 243 (1983).
4. Barton, N. G., *J. Australian Math. Soc, B25,* 287 (1984).
5. Brenner, H., *J. Colloid Interface Sci, 71,* 189 (1979).
6. Bird, R. B., Stewart, W. E., and Lightfoot, E. N., "Transport Phenomena," John Wiley, New York, 1960.
7. Chatwin, P. C., *J. Fluid Mech, 43,* 321 (1970).
8. Chatwin, P. C., *J. Fluid Mech, 48,* 689 (1971).
9. Chatwin, P. C., *J. Fluid Mech, 58,* 657 (1973).
10. Chatwin, P. C., *J. Fluid Mech, 71,* 513 (1975).
11. Chatwin, P. C., *J. Fluid Mech, 77,* 593 (1976).
12. Chatwin, P. C., *J. Fluid Mech, 80,* 33 (1977).
13. Courant, R., and Hilbert, D., "Methods of Mathematical Physics," Vol. II, Interscience Publishers, New York, 1962, (pp. 204.)
14. Crank, J., "The Mathematics of Diffusion," Oxford University Press (1956).
15. DeGance, A. E., and Johns, L. E., *Appl. Sci. Res. 34,* 189 (1978).
16. DeGance, A. E., and Johns, L. E., *Appl. Sci. Res. 34,* 227 (1978).
17. DeGance, A. E., and Johns, L. E., *AIChE J. 26,* 411, (1980).
18. Dewey, R. J. and Sullivan, P. J., *J. Fluid Mech. 125,* 203 (1982).
19. Doshi, M. R., Gill, W. N., and Subramanian, R. S., *Chem. Engng. Sci. 30,* 1467 (1975).
20. Doshi, M. R., Daiya, P. M., and Gill, W. N., *Chem. Engng. Sci. 33,* 795 (1978).
21. Fife, P. C., and Nicholes, K. R. K., *Proc. Roy. Soc. Lond. A344,* 131 (1975).
22. Gill, W. N., *Proc. Roy. Soc. Lond. A298,* 335 (1967).
23. Gill, W. N., and Sankarasubramanian, R., *Proc. Roy. Soc. Lond. A316,* 341 (1970).
24. Gill, W. N., and Sankarasubramanian, R., *Proc. Roy. Soc. Lond. A322,* 101 (1971).
25. Gill, W. N., and Sankarasubramanian, R., *Proc. Roy. Soc. Lond. A327,* 191 (1972).
26. Gill, W. N., *Chem. Engng. Sci. 30,* 1123 (1975).
27. Griffiths, A., *Proc. Phys. Soc. Lond. 23,* 190 (1911).
28. Hildebrand, F. B., "Advanced Calculus for Applications," 2nd Edition, Prentice Hall, New Jersey (1976).
29. Hunt, B., *Int. J. Heat Mass Transfer 20,* 393 (1977).
30. Johns, L. E., and DeGance, A. E., *Chem. Engng. Sci. 30,* 1065 (1975).
31. Krishnamurthy, S., and Subramanian, R. S., *Sepn. Sci. 12,* 347 (1977).
32. Lee, G. Y., "A Study of Dispersion in Open Channel Turbulent Flows," Ph.D. Thesis in Chemical Engineering, Clarkson University (1974).
33. Lighthill, M. J., *J. Inst. Math. Appl. 2,* 97 (1966).
34. Nunge, R. J., and Gill, W. N., *Ind. Engng. Chem. 61,* No. 9, 33 (1969).
35. Nunge, R. J., Lin, T. S., and Gill, W. N., *J. Fluid Mech. 51,* 363 (1972).
36. Nunge, R. J., *Atm. Environ. 8,* 969 (1974).
37. Nunge, R. J., *Atm. Environ. 8,* 984 (1974).
38. Nunge, R. J., and Subramanian, R. S., in "Dispersion and Control of Atmospheric Emissions: New Energy-Source Pollution Potential," A.I.Ch.E. Symp. Series No. 165, Ed. R. L. Byers, B. B. Crocker, and D. W. Cooper, *73,* 10 (1977).
39. Posner, M., and Gill, W. N., *AIChE J. 19,* 151 (1973).
40. Reis, J. F. G., and Lightfoot, E. N., *AIChE. J. 22,* 779 (1976).
41. Sankarasubramanian, R., and Gill, W. N., *Proc. Roy. Soc. Lond. A329,* 479 (1972).
42. Sankarasubramanian, R., and Gill, W. N., *Proc. Roy. Soc. Lond. A333,* 115 (1973); *A341,* 407 (1974).
43. Smith, R., *J. Fluid Mech. 105,* 469 (1981).
44. Smith, R., *J. Fluid Mech. 120,* 71 (1982).
45. Smith, R., *J. Fluid Mech. 129,* 347 (1983).
46. Smith, R., *J. Fluid Mech. 134,* 161 (1983).
47. Subramanian, R. S., *J. Chrom. 101,* 253 (1974).
48. Subramanian, R. S., Gill, W. N., and Marra, R. A., *Can. J.*

Chem. Engng. 52, 563 (1974).
49. Subramanian, R. S., and Gill, W. N., *Can. J. Chem. Engng. 54,* 121 (1976).
50. Subramanian, R. S., and Berhe, S., *Chem. Engng. Sci. 31,* 1005 (1976).
51. Subramanian, R. S., *Chem. Engng. Sci. 32,* 788 (1977).
52. Taylor, G. I., *Proc. Roy. Soc. Lond. A219,* 186 (1953).
53. Taylor, G. I., *Proc. Roy. Soc. Lond. A223,* 446 (1954a).
54. Taylor, G. I., *Proc. Roy. Soc. Lond. A225,* 473 (1954b).
55. Tseng, C. M., and Besant, R. W., *Proc. Roy. Soc. Lond. A317,* 91 (1970).
56. Tseng, C. M., and Besant, R. W., *Int. J. Heat Mass Transfer, 15,* 203 (1972).
57. Ultman, J. S., and Blatman, H. S., *AIChE. J., 23,* 169 (1977).
58. Yu, J. S., *J. App. Mech.,* Vol. 43, *Trans. ASME 98,* Series E, 537 (1976).

STUDY PROBLEMS

1. How is miscible dispersion different from molecular diffusion?
2. Why does a uniform velocity field cause no axial dispersion?
3. Does the velocity profile have to be parabolic to cause axial dispersion? For instance, would you see axial dispersion with a linear velocity profile which occurs in simple Couette flow?
4. What was Taylor's principal contribution in his 1953 paper?
5. What is the difference between the Taylor dispersion coefficient and the Taylor-Aris dispersion coefficient?
6. Does diffusion help or hinder axial dispersion in flow systems?

HOMEWORK PROBLEMS

1. A capillary has a diameter of 4 mm and has water flowing through it at an average velocity of 50 mm/s. What would be the Taylor dispersion coefficient k for potassium permanganate in this system? The diffusivity of permanganate at room temperature is $D \approx 1.5 \times 10^{-9}$ m²/s. Compare the value of k to that of the molecular diffusivity. Is the correction for axial molecular diffusion significant in this case?
2. Taylor's dispersion equation also is often written in the following form:

$$\frac{\partial C_m}{\partial t} + v_m \frac{\partial C_m}{\partial x} = k \frac{\partial^2 C_m}{\partial x^2} \qquad \text{(H-1)}$$

Show that Equations H-1 and 10 (in the module) are equivalent by carrying out the transformation from the (t, x) system to the (t, x_1) system, using Equation 6 in the module.

3. Try to satisfy the initial condition in Equation 14 by using Equation 18—do you encounter any difficulty? How would you satisfy Equation 14 if $g = g(t, r)$ in Equation 18?
4. Two identical slugs of dilute potassium dichromate solution are introduced at the same instant into water flowing through two identical tubes of inside diameter 50 mm. The only difference between the systems is that in tube 1, the average velocity of the water $v_{m_1} = 40$ mm/s, whereas in tube 2, the average velocity $v_{m_2} = 80$ mm/s. A strange result is observed! The solute distribution in tube 2 moves down the tube twice as fast as that in tube 1, as expected, but it disperses in the axial direction to a *much smaller extent than in tube 1.* Explain this observation on the basis of the physical mechanisms influencing dispersion.
5. A circular tube has its inside wall coated with a thin layer of catalyst. A dilute mixture of two solutes is introduced as a slug at the inlet at time zero. Both the solutes travel down the tube and undergo dispersion in laminar flow, but only one of them reacts at the catalytic wall. Will the reacting solute travel at a different average velocity from the non-rcacting one? Will it disperse more or less in the axial direction? Why?

Module C6.3

Biomedical Examples of Transport Phenomena: Coupled Diffusion Effects

Robert H. Notter

University of Rochester
School of Medicine
Rochester, NY 14642

OBJECTIVES

At the completion of this module, the student should be able to:

1. Define the three major types of diffusion-related mass transport present in living systems, and give a brief discussion of the relevant mechanisms.
2. Define and discuss passive diffusional coupling, particularly with reference to counter-gradient transport in systems involving membranes.
3. Write equations describing the transport of binary or pseudobinary solution components through a membrane, including the relevant boundary conditions.

PREREQUISITE MATHEMATICAL SKILLS

1. Calculus through the solution of ordinary differential equation boundary value problems, and the formulation of partial differential equations.
2. Introduction to vector quantities.
3. Concept of numerical solution of ordinary differential equations.

PREREQUISITE ENGINEERING AND SCIENCE SKILLS

Understand and be able to use the concepts embodied in:

1. Fick's First Law.
2. General mass flux relations such as in Chapter 16 of Bird, Stewart and Lightfoot, *Transport Phenomena* (*14*).

INTRODUCTION

The transport of mass, heat and momentum are fundamentally necessary for the existence and survival of living systems. This is true at the cellular level all the way through the functioning of complex organ systems and whole organisms. For example, conductive, convective and radiative heat transfer are basic to body temperature regulation, and momentum transfer is, of course, involved in blood flow at all levels of the circulatory system. However, of all the transport phenomena, it is mass transfer processes which are the most pervasive and necessary for the functioning of biological systems.

In terms of diffusion-related mass transfer phenomena in living systems, three broad divisions may be defined as follows:

1. Simple or passive diffusion
2. Facilitated diffusion
3. Active transport

The first of these divisions, simple or passive diffusion, is commonly taken to include the ordinary kind of mass transfer most familiar to engineers, i.e. the movement of a substance down its concentration gradient from a region of high concentration to regions of low concentration. In the case of ionic transport, this category also includes transport under the influence of electric field gradients. However, all passive diffusion systems are characterized by mass transport caused by the chemical potential gradients of the diffusing species, without the presence of carrier molecules or diffusion coupled chemical reactions. The mass transfer problem examined in this module is included in this simple diffusion category, but with superimposed convective transport because of a non-zero system mass average velocity generated by the diffusion field.

As opposed to simple diffusion systems, the categories of facilitated diffusion and active transport necessarily involve the consideration of carrier molecules. Specifically, facilitated diffusion is typically defined as the net diffusive movement of a given chemical species at a rate faster than that accounted for by ordinary diffusion. This kind of facilitated diffusion almost always occurs by a mechanism whereby a carrier molecule of some kind reacts with a diffusing species in one region, forming a complex. This complex then moves down its concentration gradient to another region, where it releases species i. Alternatively, in some facilitated diffusion processes in cell membranes, the carrier–species i complex may undergo a conformational change which carries species i directly across the membrane. However, in facilitated diffusion in general, the carrier uptake and release reactions proceed in terms of their natural equilibrium, and energy input from other reactions is not required. Thus, in this mechanism, the diffusion of species i down its own concentration gradient in an unbound form is enhanced by the additional (passive) transport of carrier bound species i. A classic example of this kind of carrier facilitated diffusion is the transfer of O_2 through hemoglobin solutions, and this is treated in detail in later modules.

The third category of mass transport in biological systems is active transport. This kind of transport is perhaps the most important type for many biological processes, but it is also the most poorly understood. Mass transfer by this mechanism is called "active" because it requires energy input to the system, and such energy input is generally provided by coupled chemical reactions. In general, active transport involves the net movement of a substance against its chemical potential gradient in an energy requiring process. Thus, such transport involves a very complicated interplay of diffusive and reactive phenomena. Although the theories of non-equilibrium thermodynamics have proven helpful in phenomenological descriptions of active transport, the fundamental elements and mechanisms of the mass transport and energy coupling are not yet fully characterized. Active transport and its many complexities are not detailed further here. Instead, the present module shows an example of countergradient mass transfer produced by direct diffusional coupling effects without the necessity of active, energy requiring transport.

Concept of Passive Diffusional Coupling

In any diffusion system, the net movement of an individual component is dependent not only on its own concentration (chemical potential) driving force, but also on the movement of the other components of the system. This is true for several reasons. First, the concentration gradient present for one component may be directly related to the concentration gradients of other components, as for example in the case of a binary system. Also, in a more general sense, the summation of the velocities of all of the individual components of the system determines the overall system velocity, and this in turn affects the net transport of each individual component. This overall system velocity may be generated solely by the diffusional motion of the various components, and does not require any imposed bulk pressure driving force. In this case, its effects on individual component transport are described here as direct coupled diffusion effects.

One important example of such a process involves diffusive coupling in membrane transport systems where the membrane has different permeabilities to the various diffusing species. Even if such a system has no pressure gradients, and hence no pressure driven bulk flow through the membrane, there may still be a non-zero mass average velocity resulting from the sum of the diffusion velocities of the transporting species. If this overall mass average velocity is in a direction opposite to the concentration (chemical potential) gradient driven diffusion of a given species i, it can, in principle, overcome the concentration driving force and result in a net flux of component i against its gradient. Such a mechanism requires no outside energy expenditure or membrane bound carrier molecule, and is due entirely to diffusive coupling rather than to any kind of active transport.

For biological membranes, the phenomenon of countergradient transport by diffusional coupling is not of the widespread significance of energy requiring active transport. However, diffusional coupling effects can be present in certain cases, particularly when the transport of molecular species of widely differing permeability is involved. When this occurs in biological systems, water is generally the solvent of high permeability that is responsible for the overall bulk flow. This transport process is generally classed under the heading of "solvent drag" by physiologists. The designation solvent drag, as used by life scientists, is not always identical to the pure diffusional coupling previously described. This is because the flux of solvent (water) that leads to the non-zero system mass average velocity in solvent drag can occur from a variety of causes, including hydrostatic pressure differences. The net effect, however, is essentially the same, in that transport by diffusion and by the non-zero system mass average velocity must be taken into account.

The phenomenon of solvent drag has received attention since the early work of Ussing (*1*), Anderson and Ussing (*2*) and Rosenberg and Wilbrandt (*3*), among others, in the 1950's. Physiology texts such as Mountcastle (*4*) discuss this phenomenon, as do a variety of books involving biological system transport and membranes (e.g., *5*, *6*) Examples of experimental studies of solvent drag in living systems include Froemter, et al. (*7*), concerning kidney tubule transport and Ochsenfahrt and Winne (*8*) on intestinal transport. Theoretical considerations have also been addressed by a number of authors including Stender, et al (*9*), Galey and Van Bruggen (*10*) and Patlak and Rapoport (*11*).

This module will consider an example wherein coupled diffusion effects alone lead to countergradient mass transfer of solute across a membrane, as described by Notter, et al (*12*). Their analysis was motivated largely by the experiments of Min, et al (*13*) on the diffusive transport of ethanol and water through a cellophane membrane. In this well-defined diffusion system, Min, et al found that countergradient mass transfer of ethanol through cellophane could be generated under some conditions. Since energy requiring active transport could not be present in the cellophane-ethanol-water system, and since bulk pressure and temperature driving forces were not present, Min, et al suggested that passive coupled diffusion effects were responsible for the observed countergradient mass flux. This interpretation was given quantitative verification in Reference (*12*) from a theoretical analysis based on a continuum theory Fick's Law approach, and from further experiments on the water-ethanol-cellophane membrane system.

The Fick's Law approach has the advantage of defining mass fluxes in a way that is open to macroscopic experimental investigation and physical interpretation, and has been verified in terms of accuracy for the description of bulk phase mass transfer phenomena in an extremely wide variety of well defined physical systems. Moreover, this kind of analysis has been taught to a broad spectrum of engineers and scientists, particularly since the introduction of the pioneering text on transport phenomena by Bird, Stewart and Lightfoot (*14*) in 1960. Consequently, the theoretical analysis on which the remainder of this module is based utilizes the Fick's Law approach for the description of diffusional mass transport.*

Theoretical Analysis of Coupled Diffusion

Much of the following theoretical description is taken from the work of Notter, et al (*12*). To display the dominant role of passive diffusional coupling in some situations, these authors considered an ideal system of an infinite membrane separating two binary solutions, as shown in Figure 1. For simplicity, the case of non-ionic diffusing components is examined here. This constraint can be relaxed for ion transport by the inclusion of an electric field gradient term in the overall chemical potential driving force for diffusion (e.g., *14, 15*). However, the case of non-ionic diffusion is sufficient to display the relevant principles of transport by diffusion and by the non-zero system mass average velocity.

In Figure 1, the two non-ionic bulk solution components are designated by *A* and *B*, and the membrane material is assumed to be homogeneous and denoted as component *M*. The membrane surfaces are defined by positions $z = \pm l$. Very far from the membrane surfaces, at positions $z = -L_1$ and $z = +L_2$, the bulk phase concentrations are fixed at mass fraction $\omega_A = \omega_{A1}$ and $\omega_A = \omega_{A2}$ respectively. If ω_{A1} and ω_{A2} are different in magnitude, and if the temperature and pressure on both sides of the membrane are fixed and equal, then mass transfer through the membrane will be initiated by the $\omega_{A1} - \omega_{A2}$ concentration difference. The equations describing the mass concentrations and mass fluxes of components *A* and *B* at steady state can then be developed.

In this analysis, the membrane is assumed to be fixed in space so that the velocity of the membrane itself is zero, i.e. membrane fixed coordinates are employed. However, the mass average velocity **v** of the system in terms of the transporting species *A* and *B* is not necessarily zero, and is given for a binary system by (*14*):

$$\mathbf{v} = \frac{1}{\rho}(\rho_A \mathbf{v}_A + \rho_B \mathbf{v}_B) \tag{1}$$

In Equation 1, ρ is the density of the system, ρ_A and ρ_B are the component densities, $\mathbf{v}_A$ and $\mathbf{v}_B$ are the component velocities and **v** is the system mass average velocity. It is the existence of this system mass average velocity **v** that permits the generation of countergradient mass transfer, even in a passive diffusion system for *A* and *B*.

In order to determine the conditions under which countergradient transport of *A* or *B* could occur in the membrane transport system of Figure 1, it is necessary to write explicit expressions for the total mass fluxes of these components. The procedure for accomplishing this is set forth in general in Chapter 16 of Bird, et al (*14*). The resulting expressions are as follows for the system of Figure 1, with the region of applicability denoted by superscripts I, II or III:

*In the physiological literature, descriptions of membrane transport processes (including solvent drag effects) generally use a different formalism based on irreversible thermodynamics. An introduction to this approach is available in a variety of books (eg. *15*).

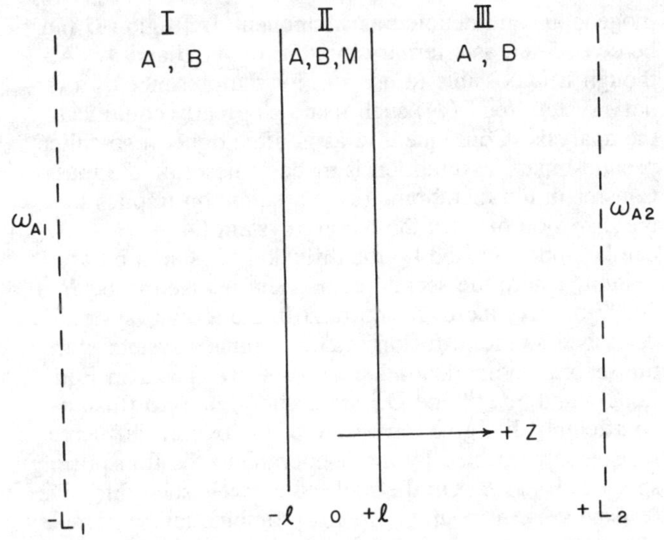

Figure 1. Coordinate system for membrane transport analysis (*12*).

Bulk Binary Solutions (Regions I and III)

$$n_A^{\mathrm{I}} = -\rho^{\mathrm{I}}\mathbf{D}_{AB}\frac{d\omega_A^{\mathrm{I}}}{dz} + \omega_A^{\mathrm{I}}(n_A^{\mathrm{I}} + n_B^{I}) \tag{2}$$

$$n_A^{\mathrm{III}} = -\rho^{\mathrm{III}}\mathbf{D}_{AB}^{\mathrm{III}}\frac{d\omega_A^{\mathrm{III}}}{dz} + \omega_A^{\mathrm{III}}(n_A^{\mathrm{III}} + n_B^{\mathrm{III}}) \tag{3}$$

Membrane (Region II)

$$n_A^{\mathrm{II}} = -\rho^{\mathrm{II}}\mathbf{D}_A^{\mathrm{II}}\frac{d\omega_A^{\mathrm{II}}}{dz} + \omega_A^{\mathrm{II}}(n_A^{\mathrm{II}} + n_B^{\mathrm{II}}) \tag{4}$$

and

$$n_B^{\mathrm{II}} = -\rho^{\mathrm{II}}\mathbf{D}_B^{\mathrm{II}}\frac{d\omega_B^{\mathrm{II}}}{dz} + \omega_B^{\mathrm{II}}(n_A^{\mathrm{II}} + n_B^{\mathrm{II}}) \tag{5}$$

In Equations 2 through 5, n_A and n_B are the total mass fluxes of components *A* and *B* with respect to the membrane fixed coordinates of Figure 1. Although these quantities are in general vectors, the membrane transport system under discussion gives rise only to mass transfer in the *z*-direction (normal to the membrane). Thus, in Equations 2 through 5 and in subsequent equations, vector notation has been dropped and all descriptions refer to the *z*-component of mass flux. Also in these equations, ω_i is the mass fraction of component *i* (ρ_i/ρ), and the various **D** quantities are diffusion coefficients as discussed next.

In Equations 2 and 3, $\mathbf{D}_{AB}$ is the normal Fick's Law binary diffusion coefficient (*14*). However, in the membrane region II, the situation is more complicated. Specifically, even if the membrane material is assumed homogeneous and denoted as component M, region II must be considered as a ternary mixture of A, B and M. Although it is possible to account for the presence of a ternary system (eg. *14*), such relations greatly complicate the analysis. Consequently, as is often done, a so-called pseudo-binary assumption is made to describe the mass transfer in the membrane. This assumption implies that the diffusion of A in the ternary system $(A + B + M)$ can be approximated by the diffusion of A in a binary system, where the second component is taken to be $B + M$. Similarly, the diffusion of B in the ternary system is described as the diffusion of B in a binary system with the second component taken as $A + M$. Thus, in Equations 4 and 5, $\mathbf{D}_A^{II}$ and $\mathbf{D}_B^{II}$ are pseudo-binary diffusion coefficients that give a measure of the overall diffusive resistance presented by the membrane to the transporting species A and B. On the molecular level, such diffusive resistance is, of course, due to a combination of complex phenomena involving intermolecular forces and collisional interactions between A, B and M. Moreover, if microscopic pores or channels are present in the membrane, stearic hindrance effects will be present. All of these complex effects are lumped together in the Fick's Law approach, and expressed in terms of effective pseudo-binary diffusion coefficients $\mathbf{D}_A^{II}$ and $\mathbf{D}_B^{II}$.

The form of Equations 2 through 5 shows explicitly the mechanism of what has been described here as passive diffusional coupling effects. Specifically, the mass flux (n_A or n_B) in these equations is made up of two contributions—a diffusional mass flux generated by the concentration gradient of the given component, together with a term expressing the transport of that component due to the total mass flux of the mixture. Again, this latter mass flux is a result of the overall system mass average velocity, and in the problem under investigation here, it is generated entirely by the diffusion process itself, since no bulk pressure gradients are present.

For steady state, one-dimensional mass transfer through a system of uniform cross-section as in Figure 1, overall and component mass balances show that the individual fluxes n_A and n_B, and the total flux $K = n_A + n_B$, are constants. Equations 2–5 can then be integrated with appropriate boundary conditions and constraints to give Equations 6–9 (*12*).*

Total flux $K \neq 0$

$$\frac{n_A}{K}=\frac{\omega_{A1}e^{\left[\frac{K}{\mathbf{D}}(L_1-l)+\frac{2Kl}{D_A}\right]}-\omega_{A2}e^{\frac{K}{\mathbf{D}}(l-L_2)}}{\left\{\left(1-\frac{1}{k_A}\right)\left(1-e^{\frac{2Kl}{D_A}}\right)+e^{\left[\frac{K}{\mathbf{D}}(L_1-l)+\frac{2Kl}{D_A}\right]}-e^{\frac{K}{\mathbf{D}}(l-L_2)}\right\}} \tag{6}$$

and

$$\frac{n_A}{K}=\frac{\left\{\left(1-\frac{1}{k_B}\right)\left(1-e^{\frac{2Kl}{D_B}}\right)+\omega_{A1}e^{\left[\frac{K}{\mathbf{D}}(L_1-l)+\frac{2Kl}{D_B}\right]}-\omega_{A2}e^{\frac{K}{\mathbf{D}}(l-L_2)}\right\}}{\left\{\left(1-\frac{1}{k_B}\right)\left(1-e^{\frac{2Kl}{D_B}}\right)+e^{\left[\frac{K}{\mathbf{D}}(L_1-l)+\frac{2Kl}{D_B}\right]}-e^{\frac{K}{\mathbf{D}}(l-L_2)}\right\}} \tag{7}$$

Total flux $K = 0$

$$n_A=-n_B=\frac{k_A(\omega_{A1}-\omega_{A2})}{k_A(L_1+L_2-2l)/\mathbf{D}+2l/D_A} \tag{8}$$

subject to the constraint,

$$\frac{k_A D_A}{k_B D_B}=1 \quad (\text{for } K=0) \tag{9}$$

In Equations 6–9, the superscripts I, II and III have been dropped from n_A and n_B because of the equality of species i mass flux in all three regions at steady state. The parameters $\mathbf{D}$, D_A and D_B are density–diffusivity products that are assumed constant with position z, i.e.

$$\rho^{I}\mathbf{D}_{AB}^{I}=\rho^{III}\mathbf{D}_{AB}^{III}=\mathbf{D} \tag{10}$$

$$\rho^{II}\mathbf{D}_A^{II}=D_A \tag{11}$$

$$\rho^{II}\mathbf{D}_B^{II}=D_B \tag{12}$$

where $\mathbf{D}$, D_A and D_B are constants. The approximations in Equations 10–12 are sufficiently accurate not to limit the major scope of the analysis here. Finally, in Equations 6–9, the k_A and k_B parameters are distribution coefficients that express the solubility of component i in the membrane. Thus, all solubility effects at the two mem-

*Equations 6–9 look relatively complex, but their derivation is primarily algebraic once Equations 2–5 have been formulated. This will become more evident when a related, but simplified, system is considered in problem 2 at the end of this module.

brane-binary solution interfaces are lumped together and expressed in terms of a macroscopic solubility parameter defined by

$$\omega_i \begin{pmatrix} \text{in membrane} \\ \text{at interface} \end{pmatrix} = k_i \omega_i \begin{pmatrix} \text{in bulk solution} \\ \text{at interface} \end{pmatrix} \quad (13)$$

In summary, Equations 6–9 express the steady state mass flux results for the membrane transport system of Figure 1. Unfortunately, Equations 6 and 7, which give n_A and n_B for the case of total flux $K \neq 0$ are transcendental in nature. Consequently, direct analytical solution of these equations for n_A and n_B is not possible.

The mass flux solutions in Equations 6 and 7 were solved numerically by Notter, et al (*12*) to obtain individual values of n_A and n_B. Equations 6 and 7 were combined into a single nonlinear equation of the form $f(K) = 0$. This equation was then solved by a Newton-Raphson technique. In these solutions, many of the parameters in Equations 6 and 7 were held constant at values representative of the experimental system investigated by these authors. Specific values are given in Table 1 (see Figure 1 for orientation).

For the purposes of this module, only one set of numerical solutions is given here. Specifically, Figures 2 and 3 show the calculated mass fluxes n_A and n_B as functions of ω_{A2} for a fixed value of $\omega_{A1} = 0.3$. Each figure is a family of flux curves at various $D_B/D_A \geq 1$ ratios, so that A is taken as the least permeable component through the membrane. The most significant result displayed on Figures 2 and 3 is that countergradient transport of the least permeable component A is indeed predicted under some conditions. Specifically, on Figure 2 for $D_B/D_A = 10$, the direction of net transfer of component A is in the $-z$ direction ($n_A < 0$) for $0.1 \lesssim \omega_{A2} < 0.3$ and in the $+z$ direction ($n_A > 0$) for $\omega_{A2} > 0.3$, all for ω_{A1} fixed at 0.3. However, reference to Figure 1 shows that the direction of the concentration gradient of A is opposite to these flux directions for $\omega_{A2} \gtrsim 0.1$. This countergradient transport of A is even more pronounced along the $D_B/D_A = 100$ curve of Figure 2, where n_A is against the gradient for essentially all values of $\omega_{A2} > 0$.

These countergradient transport results are explained by Notter, et al as follows. For any flux curve on Figures 2 or 3 with $D_B/D_A > 1$, the membrane is more permeable to component B than to component A.* Thus for $D_B/D_A > 1$, there is a net mass average velocity imposed on the system by the diffusion field, and diffusion generated convection effects are then superimposed on the diffusive transport. This fact does not affect the direction of net transport for the more permeable species B, since its greater flux is in effect generating the convection. Thus, the n_B values on Figure 3 are always in the direction of the component B concentration gradient. However, the net direction of transport for the less permeable component A is affected if enough A is carried

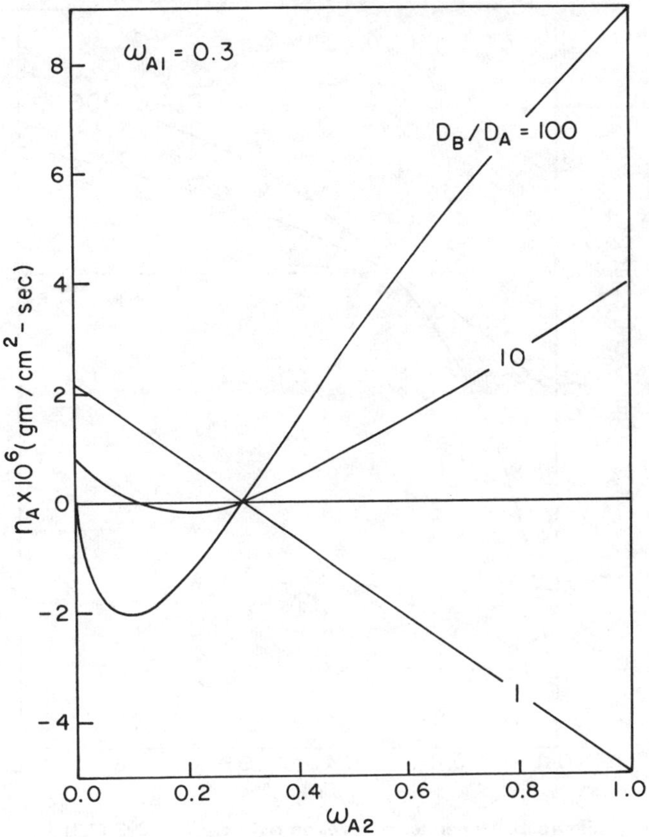

Figure 2. Dependence of n_A on ω_{A2} at ω_{A1} = 0.3 (*12*).

Table 1. Parameter values in the numerical solutions of Notter, Tam and Min (*12*).

l	=	0.002 cm
$L_1 = L_2$	=	0.20 cm
$k_A = k_B$	=	0.4
$\mathbf{D}$	=	10^{-5} gm/cm − s
D_A	=	10^{-7} gm/cm − s
D_B	=	10^{-7}, 10^{-6}, 10^{-5} gm/cm − s
ω_{A1}	=	0, 0.3, 0.6
ω_{A2}	=	0 − 1.0

by the non-zero mass average velocity to outweigh the diffusive transport of A in the opposite direction.

In the absence of any system mass average velocity (i.e. total flux $K = n_A + n_B = 0$), no countergradient transport can be generated by the pure diffusion phenomena considered here. This zero total mass average velocity case is represented on Figures 2 and 3 by the $D_B/D_A = 1$ line, since the given solutions are for the case $k_A = k_B$. Thus, the $D_B/D_A = 1$ line represents simple counter diffusion with the fluxes of A and B always in the direction of their imposed concentration gradient driving force.

In conclusion, the results presented in this module show that diffusion phenomena, particularly in the presence of a membrane, can lead to somewhat unexpected results. In diffusion through membranes, the general bias that the net mass transfer of a specific component will follow its chemical potential gradient may not always

*In general permeability is a function of distribution coefficient k_i as well as diffusivity, but in these solutions k_A and k_B have been assumed equal.

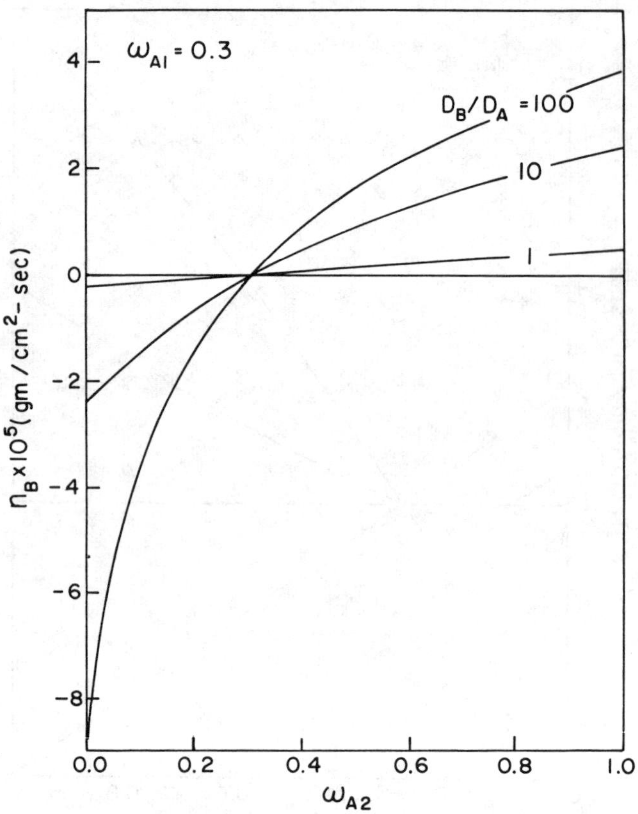

Figure 3. Dependence of n_B on ω_{A2} at ω_{A1} = 0.3 (*12*).

hold, if conditions permit a non-zero system mass average velocity. Conversely, all countergradient transport through membranes is not "active" transport, requiring energy input and specific carrier molecules, although this mechanism is certainly dominant in biological systems.

NOMENCLATURE

A, B	=	non-ionic species A, species B in diffusion system (Figure 1)
$\mathbf{D}_{AB}$	=	binary diffusion coefficient of A, B
$\mathbf{D}$	=	$\rho \mathbf{D}_{AB}$
$\mathbf{D}_A$, $\mathbf{D}_B$	=	effective pseudobinary diffusion coefficients of A, B in the membrane
D_A, D_B	=	$\rho \mathbf{D}_A$, $\rho \mathbf{D}_B$
k_A, k_B	=	distribution coefficients of A, B in membrane
K	=	total flux, $n_A + n_B$, in z-direction
l	=	one half of membrane thickness (Figure 1)
L_1, L_2	=	positions of fixed bulk concentration far from membrane (Figure 1)
M	=	membrane material
n_i	=	mass flux of species i relative to membrane fixed coordinates (in z-direction)
$\mathbf{v}$	=	vector mass average velocity
z	=	direction normal to membrane

Greek letters

ρ_i	=	density of species i
ρ	=	overall density = $\Sigma \rho_i$
ω_i	=	mass fraction of i = ρ_i/ρ
ω_{A1}, ω_{A2}	=	fixed mass fractions of A far from membrane (Figure 1)

Subscripts

A, B	=	refer to species A, species B

Superscripts

I, III	:	Bulk phase regions adjacent to membrane (Figure 1)
II	:	Membrane region (Figure 1)

LITERATURE CITED

1. Ussing, H. H., "Advances in Enzymology," *13*, 21–65 (1952).
2. Andersen, B. and H. H. Ussing, *Acta Physiol. Scand.*, *39*, 228–239 (1957).
3. Rosenberg, T. and W. Wilbrandt, *J. Gen. Physiol.*, *41*, 289–296 (1957).
4. Mountcastle, V. B. Ed., "Medical Physiology," 14th ed., C. V. Mosby, St. Louis, MO. (1980).
5. Andreoli, T. E., Hoffman, J. F., Fanestil, D. D. and S. G. Schultz, Eds., "Physiology of Membrane Disorders," 2nd ed., Plenum Medical Book Co., New York (1986).
6. Staub, N. C. and A. E. Taylor, Eds., "Edema," Raven Press, New York (1984).
7. Froemter, E., Rumrich, G. and K. J. Ullrich, *Pfluegers Arch. Eur. J. Physiol.*, *343*, 189–220 (1973).
8. Ochsenfahrt, H. and D. Winne, *Naunyn-Schmiedeberg's Arch. Pharmacol.*, *279*, 133–152 (1973). Ibid, *281*, 175–217 (1974).
9. Stender, S., Kristensen, K. and J. Skadhauge, *Memb. Biol.*, *11*, 377–398 (1973).
10. Galey, W. R. and J. T. Van Bruggen, *J. Gen. Physiol.*, *55*, 220–242 (1970).
11. Patlak, C. S. and S. I. Rapoport, *J. Gen. Physiol.*, *57*, 113–124 (1971).
12. Notter, R. H., Tam, Y. M., and S. Min, *A.I.ChE.J.*, *25*, 469–477 (1979).
13. Min, S., Duda, J. L., Notter, R. H. and J. S. Vrentas, *AIChE Journal*, *22*, 175–182 (1976).
14. Bird, R. B., Stewart, W. E., and E. N. Lightfoot, "Transport Phenomena," John Wiley & Sons, New York (1960).
15. Schultz, S. G., "Basic Principles of Membrane Transport," Cambridge University Press, Cambridge, U.K. (1980).

STUDY PROBLEMS

1. Would you expect diffusional coupling or solvent drag effects of the kind discussed here to be more important physiologically in transport into individual cells or in transport across tissue membranes? Can you name some specific human organ systems where the effect might be relevant? What living system component should be the most relevant for the role of the more permeable component B?

2. If $D_B = D_A$, could countergradient transport of one of the components A or B still be predicted for the system of Figure 1? Why?

3. If $\omega_{A2} = \omega_{A1}$, are Equations 6 and 7 still applicable?

4. What is the reason for the minimums in the $D_B/D_A > 1$ curves of Figure 2?

HOMEWORK PROBLEMS

1. This module analyzed a three region membrane transport system at steady state and derived two equations

for n_A and $K = n_A + n_B$, given by Equations 6 and 7.

a) Rearrange these equations to develop two constraints that must hold for n_A and n_B to be in the same direction (i.e. countergradient transport for one component) in the form:

$$\frac{n_B}{n_A} > 0 \quad \text{if} \quad \frac{[\ \ ]}{[\ \ ]} > 1$$

$$\text{and} \quad \frac{[\ \ ]}{[\ \ ]} > 1$$

b) Values for the various parameters in these constraints are given as follows:

$L_1 = L_2 = 0.2$ cm
$l = 0.002$ cm
$\mathbf{D} = 10^{-5}$ gm/cm s
$D_A = 10^{-7}$ gm/cm s
$D_B = 10^{-6}$ gm/cm s
$k_A = k_B = 0.4$
$K = 10^{-5}$ gm/cm^2 – s

If $\omega_{A1} = 0.12$, find the approximate range of values of ω_{A2} for which $n_B/n_A > 0$, i.e., find the range of ω_{A2} for which countergradient transport of A occurs.

c) Discuss *briefly* the *limitations* of this approach of finding ranges of concentration where countergradient transport should exist, as opposed to the numerical solutions for n_A and n_B given in the module.

2. Instead of the three region problem considered in Homework Problem 1, a simpler case is the corresponding one region problem of an isolated membrane as shown in Figure 4. All the mass fractions ω_i for this simplified analysis, including the boundary concentrations ω_{i1} and ω_{i2}, are *in* the membrane. Steady state is again assumed. Analyze this problem in an analogous fashion to the more complex case, to obtain an expression or expressions for the fluxes n_A and n_B. Are the results subject to the same transcendental form as those found for the three region case?

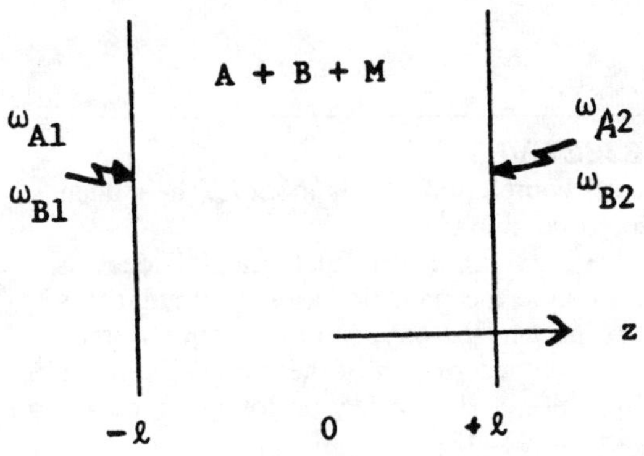

Figure 4.

Module C6.4

Biomedical Examples of Transport Phenomena—Facilitated Diffusion

Robert H. Notter

University of Rochester
School of Medicine
Rochester, N.Y. 14642

OBJECTIVES

At the completion of this module, the student should be able to:

1. Define facilitated diffusion and discuss its possible mechanistic bases, particularly with respect to the oxygen-hemoglobin system.
2. Give a description of the oxygen-hemoglobin reaction and its effect on total oxygen solubility in blood.
3. Write equations describing oxygen diffusion facilitation by hemoglobin in both the reaction equilibrium and non-equilibrium cases.
4. Discuss the probable physiologic importance of facilitated diffusion of oxygen in the red blood cell *in vivo*.

PREREQUISITE MATHEMATICAL SKILLS

1. Calculus through the solution of ordinary differential equation boundary value problems, and the formulation and meaning of partial differential equations.
2. Concept of numerical solution of ordinary differential equations.

PREREQUISITE ENGINEERING AND SCIENCE SKILLS

1. Understand and be able to use the concepts embodied in Fick's First Law.
2. General mass flux relations such as in Reference (*15*), Chapter 16.
3. Introductory kinetics in terms of equilibrium constant and rate constant.

INTRODUCTION

As outlined in a companion module (C6.3), diffusive mass transfer phenomena in biological systems are not confined solely to the simple or passive diffusion systems most familiar to engineers. One of the more interesting kinds of such transport is the phenomenon of facilitated diffusion, and this forms the subject of the present module. As defined by Dorland's Medical Dictionary (*1*), facilitation is "the promotion or hastening of any natural process; the reverse of inhibition." Thus, facilitated diffusion is generally taken to mean the diffusive transport of a substance at a faster rate than can be accounted for by ordinary diffusion of that substance down its own concentration (chemical potential) gradient. The mechanism by which such transport typically occurs is by the existence of carrier molecules which react with and "bind" the transporting substance, thus providing a second pathway to enhance the overall diffusion rate.

The existence of facilitated transport systems in a wide variety of biologically relevant processes has become apparent. For example, in 1964, Stein (*2*) reviewed previous research which showed facilitated diffusion systems either existing or suspected of existing in bacteria (eg. E. Coli), yeasts, mammalian red blood cells, lymph node cells, fibroblasts, placental cells, muscle cells, and intestinal cells, among others. Also, Schultz, et al (*3*) have reviewed various carrier mediated facilitated transport studies involving more than a dozen distinct reaction systems. This topic is also discussed in books dealing with biological membranes (e.g. *4*), as well as in general physiology texts such as Mountcastle (*5*).

An understanding of facilitated transport requires knowledge about the reaction kinetics of the carrier molecules involved, over and above a description of the diffusional processes occurring. There are several levels on which such understanding can be attempted. For many biological systems involving membrane bound carrier molecules, the molecular structures of the carrier molecules are not known. Similarly, the reaction mechanisms of these molecules with the diffusing component, including specific carrier conformational changes that may occur after binding, have often not been completely characterized. Thus, one level of analysis of biological facilitated diffusion processes involves descriptions in terms of generalized carrier uptake and release reactions, but where the molecular details of the carrier are not necessarily specified precisely. The facilitated diffusion of a particular substance can then be characterized, for example, by using parameters such as the maximum velocity of penetration (V_{max}) and an effective equilibrium constant (K_m) for the carrier uptake-release reaction (see

Homework Problem 3). This is one approach considered in detail by Stein (*2*).

A second, more fundamental approach to characterizing facilitated diffusion systems is possible, if the carrier system is well-defined. If the actual reaction kinetics are known, this knowledge can be combined with diffusion theory to provide a more complete understanding of the overall process. This approach will be used here to provide at least an introductory understanding of the basic principles of carrier facilitated diffusion, and making it possible to contend with the formidable biological literature. To accomplish this, let us analyze a well-studied and relatively well-defined system, the facilitated diffusion of oxygen through hemoglobin solutions.

FACILITATED DIFFUSION OF OXYGEN

Oxygen-Hemoglobin Reactions

Because of its central role in aerobic metabolism, oxygen is required for the maintenance of life in organisms ranging in size from a single cell to billions of cells. In complex organisms such as mammals, the oxygen transport area required is large, and so specialized organs such as the lung have been developed for gas exchange processes between the organism and surrounding air (a pulmonary surface area of 1 m^2/kg body weight is a rough approximation for the gas exchange area available in the lungs of humans). In addition to a large surface area, gas exchange organs such as the mammalian lung require something further in order to be effective; they need to be coupled to a circulatory system containing a fluid (eg. blood) which has the capacity to carry large amounts of oxygen to the tissues, and to remove waste products produced by tissue metabolic processes.

In mammals, oxygen is primarily carried in combination with a protein, hemoglobin (Hb), which forms the bulk of the interior of the red blood cell or erythrocyte. Oxygen is relatively insoluble in normal aqueous solution, and in blood plasma (no Hb) has a solubility of about 3×10^{-3} mL O_2/100 mL plasma/mmHg O_2 (*5, 6*).* This low physical solubility of oxygen is insufficient to maintain life in all but the most simple organisms. However, in the red blood cell, large amounts of oxygen are carried by hemoglobin in a bound form called oxyhemoglobin (HbO_2), so that the overall solubility of O_2 (bound plus free) in blood is about 20 mL O_2/100 mL blood, based on a hemoglobin content of 15 gm Hb/100 mL blood and on an O_2 partial pressure of 100 mmHg as typical for the lung alveolar air sacs at sea level (*5, 6*).

The detailed kinetics of the various oxygen-hemoglobin reactions are complex. The reaction equilibrium is known to depend not only on oxygen and hemoglobin concentration but also on a variety of factors such as temperature, pH, and the concentration of carbon dioxide (CO_2) and other molecules such as 2,3 diphosphoglycerate (2,3-DPG). Moreover, to complicate the picture further, there are many different types of hemoglobin based on amino acid sequence and composition, including differences between the hemoglobin present in red blood cells from newborns compared to adults. Many of the dependencies of the O_2-Hb-HbO_2 reaction system on physicochemical variables can be explained by the fact that they allow optimal delivery of oxygen to body tissues under a variety of physiologic conditions (*5, 6*). This extensive topic will not be described further here, but it is necessary for our discussion to consider briefly the general characteristics of the hemoglobin-oxyhemoglobin reaction itself.

Each hemoglobin molecule actually contains four Hb groups of molecular weight approximately 17,000 (total MW 67,000). Each of these Hb groups can bind one oxygen molecule, but the affinity of binding is dependent upon how many of the four Hb groups already carry a bound oxygen. In general, the affinity becomes stronger as more oxygen molecules are bound, so that the affinity of an Hb_4 molecule to bind a fourth O_2 molecule is many times greater than for binding the first (*5–7*). For simplicity, however, it has become conventional to present the oxygen-hemoglobin reaction in a single step as:

$$Hb + O_2 \underset{k}{\overset{k'}{\rightleftharpoons}} HbO_2 \tag{1}$$

where k' and k are rate constants for the forward and backward reactions, respectively, and Hb is taken to have a molecular weight of about 17,000. In essence, this simplification ignores oxygen binding affinity differences for Hb subunits and smooths everything out into a single reaction relation. The rate constants k and k' are then determined empirically.

In our analysis Equation 1 will be used in two situations. One is in terms of the non-equilibrium case where consideration of both rate constants k and k' will be required; the other is for the equilibrium case, where only a single parameter is required to specify the reaction. In this latter instance, almost all physiologists define the equilibrium of Equation 1 in terms of a saturation parameter:

$$S = \frac{C_{HbO_2}}{C_{Hb} + C_{HbO_2}} \tag{2}$$

Thus, in Equation 2, S represents the fraction of hemoglobin that is saturated with oxygen at equilibrium for the given conditions of temperature, pressure, pH, oxygen and carbon dioxide partial pressures p_{O_2} and p_{CO_2}, etc. Under normal physiological conditions in man, hemoglobin is about 97% saturated as oxyhemoglobin at a p_{O_2} value of 100 mm Hg.

Many different semi-empirical expressions for S are available (*5-10*). One relatively simple expression for S is the Hill equation:

$$S = \frac{K(p_{O_2})^n}{1 + K(p_{O_2})^n} \tag{3}$$

* In this module, typical biological solubility units are employed. Thus, the units X mL O_2/100 mL blood/mm Hg O_2 imply that at standard temperature and pressure (STP), an equivalent of X mL O_2 are dissolved in 100 mL of blood for every mm Hg of O_2 partial pressure over the solution. Since 22,400 mL of O_2 form a mole of O_2 at STP, these concentrations can easily be converted to the more common engineering units of gm O_2/L.

The theoretical assumptions which originally led to Equation 3 are not correct (eg. *7, 8*), but it still has utility as an empirical relation for the equilibrium saturation S (*9*). In this case, K and n are found from fitting S to experimental data for the relevant oxyhemoglobin dissociation curve. Both K and n vary according to temperature, pH, ionic strength, 2,3-DPG concentration and other variables (*5–10*).

Mechanisms of Oxygen Diffusion Facilitation by Hemoglobin

A detailed review of much of the experimental work on O_2 facilitation by hemoglobin is given by Kreuzer (*11*), who also discusses qualitatively the various theoretical approaches used to describe the effect. A typical experimental system used to measure O_2 facilitated diffusion through hemoglobin solutions is shown in Figure 1. Oxygen or nitrogen is constrained to diffuse through a hemoglobin solution supported in a millipore filter*. The flux of nitrogen (N_2) diffusing through the solution under the influence of a partial pressure gradient $p_1 - p_2$ is taken as representative of ordinary or passive diffusion, since Hb has a negligible affinity for N_2. Because the ordinary diffusion coefficients of N_2 and O_2 in solution are very similar, any oxygen flux over the value found for nitrogen shows that facilitated diffusion is taking place for this species. This is indeed found to be the case for *in vitro* diffusion experiments under a variety of conditions (e.g. different O_2 and Hb concentrations, different solution thicknesses l, etc.) (*9, 11*).

Two major mechanisms have been proposed for the

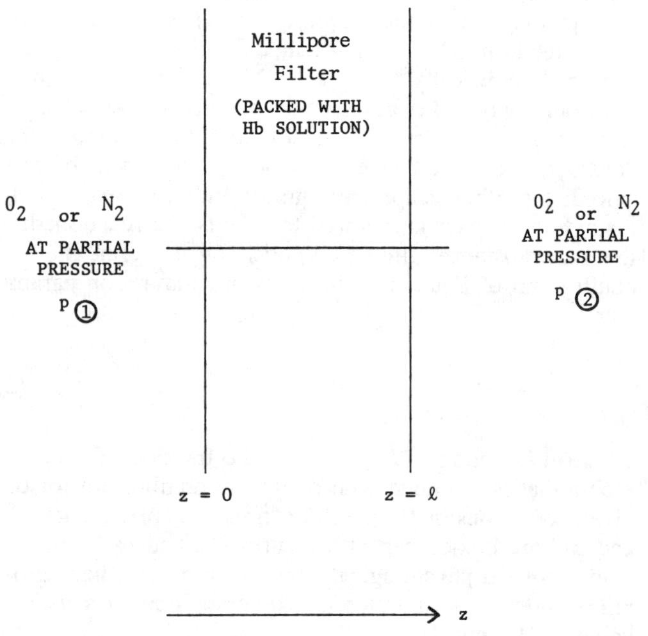

Figure 1. Typical experimental system for evaluating O_2 facilitated diffusion through hemoglobin solutions *in vitro*.

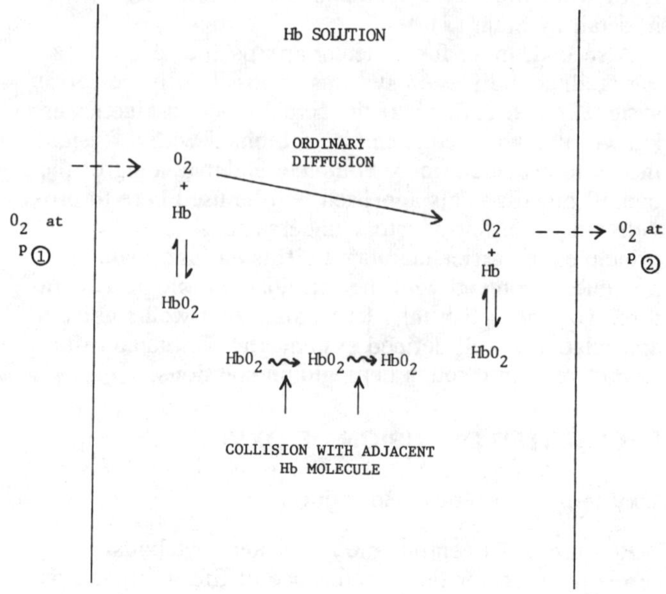

Figure 2. Collisional mechanism for O_2 facilitated diffusion through hemoglobin solutions.

facilitated diffusion of O_2 through hemoglobin, although only one is now generally accepted. The first mechanism, now discredited, is referred to as the "bucket brigade," or collisional mechanism, and is shown schematically in Figure 2. In this mechanism, oxygen is carried through the solution, both by diffusion down its own concentration gradient, as well as in combination with hemoglobin as HbO_2. However, the mechanism of O_2 transport by HbO_2 is not taken as resulting from diffusion of the HbO_2 molecules. Instead, the "bucket brigade" mechanism assumes that O_2 passes through the solution, from Hb molecule to Hb molecule, by collision. Although this mechanism is imaginative and possibly applicable to other facilitated diffusion systems, it is almost certainly not the case for the oxygen-hemoglobin system. In particular, as discussed by Kreuzer, experiments by a variety of investigators show that the oxygen-hemoglobin facilitated diffusion system does not exhibit the proper dependence on Hb concentration and temperature, or the dependence on solution thickness, for the collisional mechanism to be correct.

The second mechanism for O_2 facilitated diffusion through Hb is translational diffusion of the HbO_2 molecules themselves, as shown in Figure 3. In this mechanism, oxygen is picked up by Hb at the high concentration side of the solution until the equilibrium of Equation 1 at p_1 is reached. The HbO_2 molecules then diffuse, along with free O_2, to the low concentration side, where O_2 is given up by the natural equilibrium of Equation 1 at p_2^*.

This process of translational diffusion of HbO_2 as the facilitation mechanism seems unlikely at first glance, since the diffusion coefficient of HbO_2 is so much less

* The typical millipore filter used has pore sizes of about 0.45 microns (μ) and contains about 80% liquid space as noted by Kreuzer (*11*). Thus, the diffusion occurs essentially through a bulk liquid solution.

* Actually, since O_2 concentration changes throughout the solution in direction z, the amount of O_2 carried as HbO_2 also varies in accordance with Equation 1 throughout the solution, not just at the boundaries.

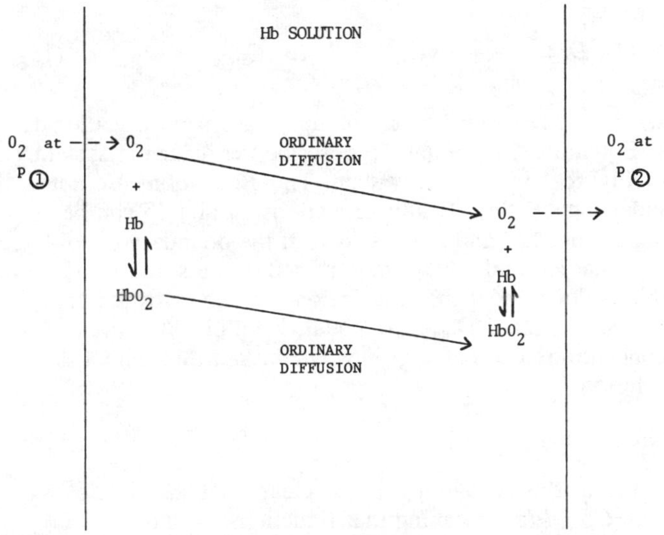

Figure 3. Translational diffusion mechanism for O_2 facilitated diffusion through hemoglobin solutions.

than that of O_2. In point, Kreuzer notes Wyman (*12*) asking "How could a fly hope to increase his rate of progress by alighting on the back of a tortoise?" However, it must be remembered that at full saturation, some two–orders–of–magnitude more O_2 can be carried as HbO_2, compared to the concentration of free O_2 in the solution. Thus, Wyman concluded: "A swarm of flies whose forward progress ... was limited by the necessity of passing through ... a barrier might well improve their situation by riding on the backs of an army of tortoises which could pass under the barrier." In other words, the much greater amount of O_2 carried as HbO_2 is enough to outweigh the slower diffusion of this large molecule.

With this mechanistic background, the quantitative description of O_2 facilitated diffusion through Hb solutions can now be given. Two approaches will be examined here, a description in terms of effective diffusion coefficient at equilibrium, and one in terms of species *i* continuity equations in the nonequilibrium case.

Description by Effective Diffusion Coefficient at Equilibrium

This approach has been followed by both Middleman (*9*) and Lightfoot (*10*) in their textbooks on transport phenomena in living systems. For the system of Figure 1, with oxygen undergoing one dimensional diffusion in the *z* direction, the relevant fluxes are defined as:

N_{O_2} = total molar flux of O_2 in direction *z*, with respect to fixed coordinates

N_{Hb} = total molar flux of Hb in direction *z*, with respect to fixed coordinates

N_{HbO_2} = total molar flux of HbO_2 in direction *z*, with respect to fixed coordinates

For this system, bulk pressure and temperature effects on diffusion are negligible. Moreover, the concentration of physically dissolved oxygen (C_{O_2}) is very dilute and its total flux N_{O_2} is small. Also, all of the Hb and HbO_2 are confined to the solution, so that $N_{Hb} + N_{HbO_2} = 0$. Consequently, the molar average velocity of the diffusion system is negligible, and the total fluxes of the various (non-ionic) components contain only concentration gradient driving force terms, i.e. (*10*):

$$N_{O_2} = -D_{O_2}\frac{dC_{O_2}}{dz} \tag{4}$$

$$N_{HbO_2} = -D_{HbO_2}\frac{dC_{HbO_2}}{dz} \tag{5}$$

In Equations 4 and 5, C_{O_2} and C_{HbO_2} are the concentrations of oxygen and oxyhemoglobin, respectively, and the total concentration of the mixture $C = \Sigma C_i$ has been assumed constant. Also, D_{O_2} and D_{HbO_2} are pseudo-binary diffusion coefficients for O_2 and HbO_2, with this approximation already discussed in Module C6.3. Finally, steady state transport is assumed, so that the various concentration terms are functions only of *z*.

If the translational diffusion mechanism of Figure 3 is applicable, the total flux of O_2, defined as $N_{Tot.}$ is given by the flux of O_2 down its own concentration gradient plus the net flux of O_2 carried as HbO_2. Thus, the addition of Equations 4 and 5 gives

$$N_{Tot.} = -\left[D_{O_2}\frac{dC_{O_2}}{dz} + D_{HbO_2}\frac{dC_{HbO_2}}{dz}\right] \tag{6}$$

Further manipulation of Equation 6 requires a knowledge of the relation between C_{O_2} and C_{HbO_2}, which comes from the oxygen-hemoglobin reaction in Equation 1. If Equation 1 is assumed to be in equilibrium, then Equation 2 gives:

$$C_{HbO_2} = S\,(C_{Hb} + C_{HbO_2}) \tag{7}$$

Equation 7 may be differentiated with respect to *z* to give an expression for the concentration gradient of HbO_2 as:

$$\frac{dC_{HbO_2}}{dz} = (C_{Hb} + C_{HbO_2})\frac{dS}{dz} \tag{8}$$

In Equation 8, it has been assumed that the total hemoglobin concentration $C_{Hb} + C_{HbO_2}$ is constant with *z*; this implies that the diffusion coefficients of Hb and HbO_2 are equal. Because of the almost uniform size and chemical properties of these large molecules, this is an excellent approximation.

Application of the chain rule of differentiation to Equation 8 can now be made. At equilibrium, the fractional saturation *S* is directly related to C_{O_2} by Equation 3, i.e. $S = S(C_{O_2})$.* Thus, it follows that

$$\frac{dS}{dz} = \frac{dS}{dC_{O_2}}\frac{dC_{O_2}}{dz} \tag{9}$$

if the other variables affecting *S* (e.g. pH) are held con-

* Actually, Equation 3 gives *S* as a function of p_{O_2}. However, $C_{O_2} = \alpha p_{O_2}$, where α is the solubility coefficient of O_2 in solution, as discussed earlier.

stant. Combining Equations 6, 8 and 9 then yields:

$$N_{Tot} = -D_{eff.} \frac{dC_{O_2}}{dz} \tag{10}$$

where $D_{eff.}$ is an effective diffusion coefficient defined by:

$$D_{eff.} = D_{O_2} + D_{HbO_2}(C_{Hb} + C_{HbO_2}) \frac{dS}{dC_{O_2}} \tag{11}$$

This steady state transport analysis at reaction equilibrium describes the facilitated diffusion of O_2 through hemoglobin solutions in terms of the effective diffusion coefficient defined by Equations 10 and 11. This coefficient is a function of position z, since it involves C_{O_2} which depends on position in the solution. From Equation 11, the effective diffusivity is made up of two parts, one which accounts for the diffusion of free (dissolved) O_2 and another which accounts for O_2 carried in bound form, as shown in Figure 3. Under appropriate conditions, this kind of analysis can have reasonable success in describing experimental O_2 facilitated diffusion data, as discussed by Middleman (*9*) and by Keller and Friedlander (*13*).

Description by Species Continuity Equations

The analysis given in the preceding section is useful for describing the facilitated diffusion of oxygen under conditions where equilibrium exists for the oxygen–hemoglobin reaction. However, in systems where oxygen concentration gradients are large (such as possible in the red blood cell *in vivo*) the equilibrium assumption is no longer valid. In this case, a more complicated, nonequilibrium analysis, with general species i continuity equations, is required. Such an analysis was carried out in a form relevant for the facilitated diffusion of O_2 by Kutchai, et al (*14*), and their work forms the basis of this section.

The general species i continuity equation for a binary or pseudo-binary system of constant diffusion coefficient and overall concentration is given, for example, in Chapter 18 of Bird, Stewart and Lightfoot (*15*) as:

$$\frac{\partial C_A}{\partial t} + \nabla \cdot C_A \mathbf{v}^* = D_{AB} \nabla^2 C_A + R_A \tag{12}$$

In Equation 12, C_A is the concentration of species A, $\mathbf{v}^*$ is the system molar average velocity, and R_A is the molar rate of production of A per volume by chemical reaction. For the oxygen–hemoglobin system of Figure 1, the molar average velocity $\mathbf{v}^*$ is small and may be neglected. Also, only transport in the z direction is considered, and steady state ($\partial/\partial t = 0$) is assumed. Under these conditions, Equation 12 may be applied to the system of Figure 1 for O_2, HbO_2 and Hb, respectively, to give (*14*):

$$O = D_{O_2} \frac{d^2 C_{O_2}}{dz^2} + kC_{HbO_2} - k' C_{O_2} C_{Hb} \tag{13}$$

$$O = D_{HbO_2} \frac{d^2 C_{HbO_2}}{dz^2} + k' C_{O_2} C_{Hb} - kC_{HbO_2} \tag{14}$$

$$O = D_{Hb} \frac{d^2 C_{Hb}}{dz^2} + kC_{HbO_2} - k' C_{O_2} C_{Hb} \tag{15}$$

where the reaction production terms are from Equation 1.

Equations 13 through 15 describe the diffusion system of Figure 1 in the steady state. These equations are not independent. Specifically, Equations 14 and 15 can be added together and integrated with the boundary condition that Hb and HbO_2 are confined to the solution. If this is done under the assumption that the diffusion coefficients D_{Hb} and D_{HbO_2} are equal, it follows that the total concentration of heme groups is constant throughout the solution, i.e.

$$C_{Hb} + C_{HbO_2} = A \tag{16}$$

From this equation, it is clear that $d^2C_{Hb}/dz^2 = -d^2C_{HbO_2}/dz^2$, meaning that Equations 14 and 15 are identical. Thus, the application of Equation 16 to Equations 13 through 15 results in a final description of the oxygen–hemoglobin system of Figure 1 by:

$$O = D_{O_2} \frac{d^2 C_{O_2}}{dz^2} + kC_{HbO_2} - k' C_{O_2}(A - C_{HbO_2}) \tag{17}$$

$$O = D_{HbO_2} \frac{d^2 C_{HbO_2}}{dz^2} - kC_{HbO_2} + k' C_{O_2}(A - C_{HbO_2}) \tag{18}$$

The boundary conditions on Equations 17 and 18, with reference to Figure 1, are—

At $z = 0$:

$$C_{O_2} = \alpha p_1 \tag{19}$$

$$\frac{dC_{HbO_2}}{dz} = 0 \tag{20}$$

At $z = l$:

$$C_{O_2} = \alpha p_2 \tag{21}$$

$$\frac{dC_{HbO_2}}{dz} = 0 \tag{22}$$

Equations 19 and 21 are statements of solubility equilibrium (with solubility coefficient α) for O_2 at partial pressures p_1 and p_2 at the solution boundaries. Also, Equations 20 and 22 follow from the fact that there can be no flux of HbO_2 across the solution boundaries at $z = 0$ and $z = l$.

Equations 17 and 18, with boundary condition Equations 19 through 22, form a set of coupled nonlinear ordinary differential equations. These equations were nondimensionalized and solved numerically by Kutchai, et al with parameters relevant for the oxygen-hemoglobin facilitated diffusion system. They expressed their results in terms of the facilitation, F, defined by

$$F = \frac{\text{Flux of } O_2 \text{ with Hb present}}{\text{Flux of } O_2 \text{ with no Hb}} - 1 \tag{23}$$

The primary results of these calculations and their agreement with experimental data, are summarized:

1. The effect of nonequilibrium for the reaction of Equation 1 is to lower the facilitation F from the equilibrium case. Nonequilibrium effects were found to be most severe near the low oxygen concentration side of the solution (ascribed by Kutchai et al to the fact that the rate constant k of the HbO_2 dissociation reaction is small compared to the association constant k'). For layer thickness $l > 25\mu$, these authors concluded that equilibrium analyses could probably be applied to the oxygen-hemoglobin system, although this limiting thickness could be greater depending on the values of p_1 and p_2 and on the amount of error that can be tolerated in a given analysis. Below $l = 25\mu$, however, errors clearly in excess of experimental uncertainty result from the equilibrium assumption.
2. Calculations based on the theoretical model showed excellent qualitative agreement with experimental data for the O_2–Hb–HbO_2 system, in terms of dependence on layer thickness l, and on oxygen partial pressures p_1 and p_2. Quantitative comparison with experiment was hampered by several factors, including the lack of precise diffusivity and rate constant data. In spite of this, however, good quantitative agreement was obtained with previous experimental studies, particularly in terms of l and p_1 variation effects.
3. Kutchai, et al solved several cases with parameters relevant for the conditions of oxygen transport into the red blood cell *in vivo*. Although several approximations were involved in this procedure, they were able to obtain some interesting and suggestive results. Human red blood cells are bi-concave discs, with the thin dimension of the cell about 1-2 microns; thus a nonequilibrium analysis is clearly required. For $l \leq 1\mu$, Kutchai, et al found that F was less than 0.35, even if the low oxygen pressure p_2 was allowed to be zero, a situation not present in the red blood cell *in vivo*. If p_2 was assumed to be 50 mmHg, the facilitation for $l = 1\mu$ was found equal to 0.19 for $p_1 = 125$ mmHg. Moreover, in the actual *in vivo* system, other molecules, such as CO_2 and 2,3-DPG, which can be bound to hemoglobin, are present; this would act to decrease F further *in vivo*. Thus, the effects of hemoglobin–mediated facilitation of O_2 transport into the red blood cell *in vivo* appear to be small.

SUMMARY

This module has shown the fundamental principles of carrier mediated facilitated diffusion systems. Most of the discussion has been in terms of the oxygen-hemoglobin system. Although this specific example of facilitated diffusion may not have a high degree of physiological importance *in vivo,* it permits a clear discussion of the analytical principles necessary for the fundamental description of non-active carrier–mediated transport. Both equilibrium and nonequilibrium reaction kinetics were examined for the oxygen–hemoglobin system, and the utility and weaknesses of each approach were delineated. Specifically, the equilibrium analysis is easy to apply, but breaks down if concentration gradients become too large. The nonequilibrium analysis, on the other hand, is more exact, but more difficult to solve and requires a more detailed knowledge of reaction rate constants.

LITERATURE CITED

1. "Dorland's Illustrated Medical Dictionary," 24th Ed., W. B. Saunders Co., Philadelphia, (1965).
2. Stein, W. D., "Recent Progress in Surface Science," Vol. I., Danielli, J. F., Pankhurst, K. G. A. and A. C. Riddiford, Eds., pp. 300-337, Academic Press, New York (1964).
3. Schultz, J. S., Goddard, J. D. and S. R. Suchdeo, *AIChE J., 20,* 417–445 (1974).
4. Andreoli, T. E., Hoffman, J. F., Fanestil, D. D. and S. G. Schultz, Eds. "Physiology of Membrane Disorders," 2nd Ed., Plenum Medical Book Co., New York (1986).
5. Mountcastle, V. B., Ed., "Medical Physiology" 14 Ed., C. V. Mosby, St. Louis (1980).
6. Comroe, J. H. "Physiology of Respiration," 2nd Ed., Year Book Medical Publishers, Chicago (1974).
7. Adair, G. S., *J. Biol. Chem., 63,* 529–545 (1925).
8. Forbes, W. H. and F. J.W. Roughton, *J. Physiol., 71,* 229–256 (1931).
9. Middleman, S., "Transport Phenomena in the Cardiovascular System," Wiley-Interscience, New York (1972).
10. Lightfoot, E. N., "Transport Phenomena and Living Systems," Wiley-Interscience, New York (1974).
11. Kreuzer, F., *Respiration Physiology, 9,* 1–30 (1970).
12. Wyman, J., *Journal Biol. Chem., 241,* 115–121 (1966).
13. Keller, K. H. and S. K. Friedlander, *J. General Physiology, 49,* 663 (1966); Ibid, 681 (1966).
14. Kutchai, H., Jacquez, J. A. and F. J. Mather, *Biophysical Journal, 10,* 38–54 (1970).
15. Bird, R. B., Stewart, W. E. and E. N. Lightfoot, "Transport Phenomena," Wiley and Sons, New York (1960).

STUDY PROBLEMS

1. What is the difference between equilibrium and steady state?

2. In terms of the behavior of S (fractional saturation in Equations 2 and 3) for oxygen-hemoglobin at equilibrium, discuss the following points with reference to facilitation F for the system of Figure 1:
   a) Would F increase proportionately if p_1 were increased from 100 mmHg O_2 to 200 mmHg O_2?
   b) If p_2 were fixed at 150 mmHg O_2 with $p_1 > p_2$, what would be the magnitude of F? Can you verify your conclusion mathematically from Equation 11?

3. If the collisional or "bucket brigade" theory of O_2 diffusional facilitation by hemoglobin were correct, what would you expect for the dependence of F on:
   a) temperature (qualitative trend)?
   b) Hb concentration?

4. Many of the approximations used in the analysis of facilitated diffusion are most accurate for dilute systems. Can a substance be dilute in molar concentration and not in weight percent concentration? Is this relevant for Hb and HbO_2?

HOMEWORK PROBLEMS

1. Follow the procedure outlined just prior to Equation 16 to verify the result that the total concentration of

heme groups is a constant at any z, i.e.

$$C_{Hb} + C_{HbO_2} = A$$

2. The effective diffusion coefficient of Equation 11 is a function of position z. However, for easy comparison with experiment, it is most convenient to use an *average* effective diffusion coefficient, $\bar{D}_{eff.}$. Construct an integral average effective diffusion coefficient for the system of Figure 1 that is related to total oxygen flux $N_{Tot.}$ by

$$N_{Tot.} = \bar{D}_{eff.} \frac{C_{O_{2}1} - C_{O_{2}2}}{l}$$

where $C_{O_2} \equiv \alpha p_{O_2}$. In terms of this average effective diffusivity, what would be the form of the facilitation F?

3. Consider the case of a carrier–mediated facilitated diffusion process wherein a substance S can *only* cross a membrane in association with a membrane bound carrier C in the form CS:

$$C + S \rightleftharpoons CS \qquad \text{(H-1)}$$

with a dissociation equilibrium constant K_m. (Note: Do not confuse the notation of substance S used in this problem with the equilibrium fractional saturation of hemoglobin S used in the text.) Here, the net transport rate N_{CS} is defined by

$$N_{CS} = D\left(\frac{C_{CS_0} - C_{CS_l}}{l}\right) \qquad \text{(H-2)}$$

where D is the diffusion coefficient for CS, and C_{CS_0} and C_{CS_l} are the concentrations of CS at the boundaries of the membrane at $z = 0$ and $z = l$. The diffusion coefficients of carrier C and carrier complex CS are assumed equal so that the total carrier concentration $C_C + C_{CS}$ = constant.

Show that Equations H-1 and H-2 imply that

$$N_{CS} = N_{CS_{max}} \left[\frac{C_{S_0}}{K_m + C_{S_0}} - \frac{C_{S_l}}{K_m + C_{S_l}}\right] \qquad \text{(H-3)}$$

where C_{S_0} and C_{S_l} are the concentrations of S at the membrane boundaries, and

$$N_{CS_{max}} = \frac{(C_C + C_{CS})}{l} D \qquad \text{(H-4)}$$

Why is the quantity in Equation H-4 identified with the maximum flux of CS? Note that it would correspond to V_{max} in the widely used biological description of facilitated transport systems mentioned earlier (*2*).

Module C6.5

Mass Transfer in Heterogeneous Media

Pieter Stroeve

University of California
Davis, CA 95616

OBJECTIVES

At the completion of this module, the student should be able to:

1. Define the factors that influence the effective diffusion coefficient of two-phase heterogeneous media, such as medium make up, physical parameters of the pure phases, volume fractions of the phases, etc.
2. Make predictions of effective diffusion coefficients for laminated media, and two-phase media where one phase is dispersed in a second continuous phase.
3. Convert the relationships for effective diffusion coefficients to relationships for other effective parameters such as thermal conductivity or electrical conductivity.
4. Work with Maxwell's, Fricke's and Bruggeman's equations. These are classical relationships in heterogeneous transport phenomena.

PREREQUISITE MATHEMATICAL SKILLS

Be able to solve:

1. Partial differential equations.
2. Legendre functions.
3. Problems in spherical coordinates.

PREREQUISITE ENGINEERING AND SCIENCE SKILLS

1. Understand Henry's Law.
2. Understand basic mass conservation equations.
3. Be familiar with other transport phenomena.

INTRODUCTION

The relationships for mass transfer taught in a basic transport phenomena course can get quite complex when you deal with real problems. In the mass transport problems encountered in the classroom, the media are usually homogeneous. But how does the problem change if you are not dealing with a homogeneous but with a heterogeneous medium? In your future work you will often deal with heterogeneous media. In the chemical industry, heterogeneous media such as emulsions, slurries, foams, dispersions, and solid mixtures are frequently encountered as process streams or materials of construction. In the bioengineering sciences one deals almost exclusively with heterogeneous systems such as cell suspensions. If a medium consists of more than one phase, how does the composition and make-up of the medium influence the effective physical properties of the medium? Physical constants may be available for pure materials but rarely for mixtures. In this module, relationships for effective physical properties of heterogeneous media are discussed. We will limit the discussion to two-phase media. In view of the many physical properties of a medium that one can be interested in, such as for example the dielectric constant, electrical conductivity, thermal conductivity, and the mass diffusivity, it's fortuitous that the different transport phenomena are analogous and that the relationships are interchangeable if appropriate modifications are made.

TYPES OF MEDIA

This module is concerned with two-phase heterogeneous media. There are numerous ways that a mixture can be composed of two phases. Figure 1 shows possible heterogeneous media. Examples of some of the media shown are: A) layered coating, or packaging material; B) cellular suspension or monodisperse latex suspension (spheres of uniform size); C) emulsion; D) soil, or filler material in a membrane; E) bacteria oriented in flowing aqueous media; F) fibers oriented in a plastic material; H) porous rock. The media shown in Figure 1 can be grouped into two categories, disperse phase in a continuous phase, and two continuous phases. Media B, C, D, E, F and G belong to the first category; media A and H belong to the second. There are even more complicated categories of two phase media; for example, two continuous phases each containing a dispersion of the other phase. Let us restrict ourselves to considering the previous two categories, and focus on some of the better known relations for the effective physical properties.

Let us look at media wherein the characteristic dimension of the particles dispersed in a continuous phase is small compared to the dimensions of the total system in which the media are contained (not for laminated media). In other words, the media possess large-scale homogeneity and isotropy, but on a smaller scale, they are heterogeneous. The literature contains many models that predict effective physical properties of two-phase media. Nearly

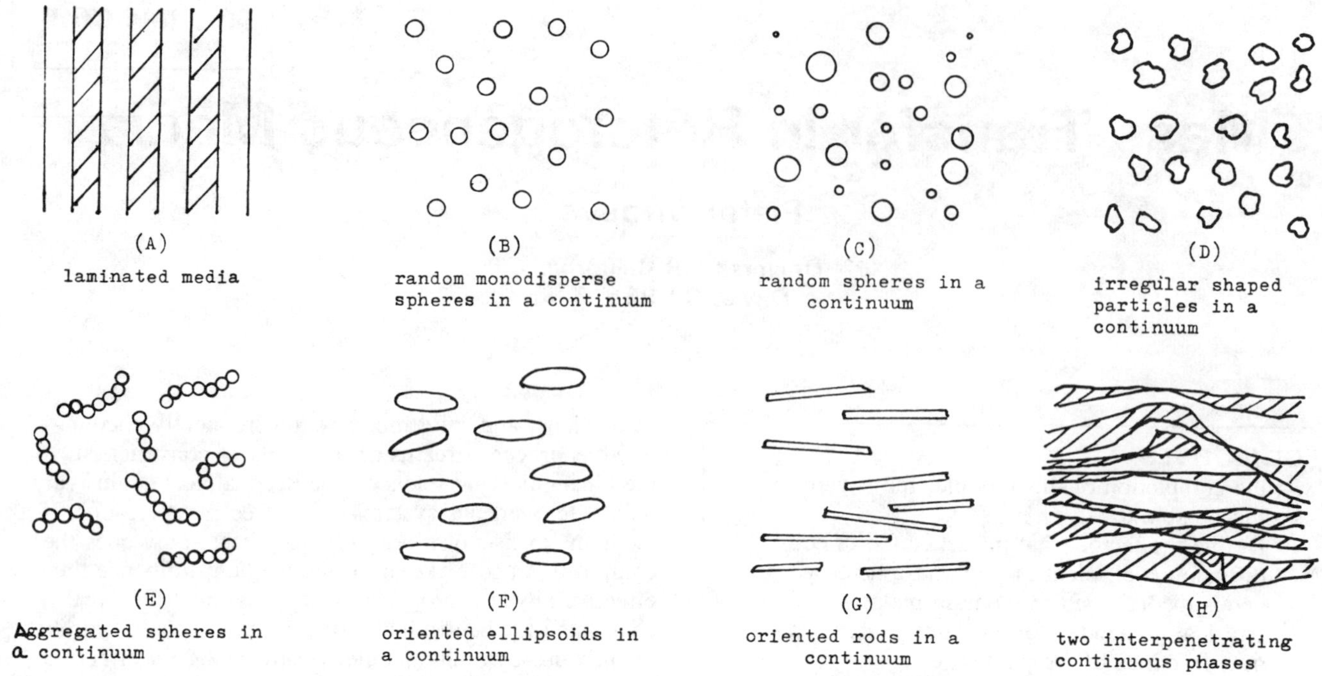

Figure 1. Examples of heterogeneous systems.

all of the analyses published to date consider the solution of the steady state problem. At steady state, many phenomena associated with heterogeneous media are expected to obey Laplace's equation,

$$\nabla^2 V = 0 \tag{1}$$

where V is the potential. In the case of the diffusion of solutes in heterogeneous media, Laplace's equation is the mass conservation equation. Several excellent reviews on the published relationships for effective physical properties are given by Barrer (*1*), and Meredith and Tobias (*2*). Examination of these reviews yields the following conclusions (emphasis on mass transfer):

i) Most of the relationships are for two-phase materials.
ii) At present, due to the complexity of the problem, no rigorous relationship is available. Exceptions are the relations derived for purely parallel or series arrangement, or infinitely dilute suspensions of spheres or spheroids.
iii) Most relations for the same type of media differ from each other in the approximations made in their development.
iv) The relations for dispersed phase in a continuum indicate that the effective relative (dimensionless) diffusivity of a species in the medium is dependent on the volume fraction of the dispersed phase, the shape of the dispersed particles, and the relative diffusion coefficients of the two phases.

$$\frac{\bar{D}K_t}{D_c} = f\left(\Phi, \text{ shape } d, \frac{D_d K_{eq.}}{D_c}\right) \tag{2}$$

where:

$\bar{D}$ = effective diffusivity of species in the medium
D_d = diffusivity of species in the dispersed phase
D_c = diffusivity of species in the continuous phase
Φ = volume fraction of the dispersed phase
$K_{eq.}$ = equilibrium distribution coefficient between dispersed phase and continuous phase, i.e.,

$$K_{eq.} = \left.\frac{C_d}{C_c}\right|_{\text{equilibrium}}$$

K_t = $1 - \Phi + \Phi K_{eq.}$
C = concentration of species
d = dispersed phase
c = continuous phase

Because there are dozens of relationships in the literature, only a few will be discussed in detail.

ONE PHASE DISPERSED IN ANOTHER CONTINUOUS PHASE

Consider diffusion through a slab of material composed of one phase dispersed in another continuous phase. For unidirectional diffusion of a species A through a layer, the flux is given by

$$J = -\bar{D}\frac{dC_t}{dx} \tag{3}$$

where $\bar{D}$ is the effective diffusivity and C_t is the concentration of A in the heterogeneous suspension. C_t can be given by

$$C_t = (1-\Phi)C_c + C_d\Phi$$
$$= (1 - \Phi + \Phi K_{eq.})C_c \tag{4}$$

Equation 3 can be rewritten in terms of the concentration of the continuous phase:

$$J = -\bar{D}K_t\frac{dC_c}{dx} \tag{5}$$

If $\bar{D}K_t$ is determined from an appropriate relationship that describes the heterogeneous medium of interest (see Equation 2), one can calculate the mass flux of the species of interest.

Maxwell (*3*) derived a relation for the simplest type of media: spherical particles of uniform size dispersed in a continuum. Maxwell first considered a single sphere in an infinite continuum where the field was unidirectional and linear at large distances from the sphere. By solving the Laplace equation for the potential in spherical coordinates inside and outside the sphere, he obtained the potential variation due to the presence of the sphere in the continuum. For a dilute suspension, he considered the suspension itself spherical and located in the continuous medium. By assuming the perturbation associated with each of the smaller spheres inside this spherical suspension to be independent of each other, he obtained an effective diffusivity for mass transfer in the suspension by equating the perturbation caused by the spherical suspension and that caused by all the small spheres. Maxwell obtained

$$\bar{D}K_t = D_c \frac{2D_c + D_d K_{eq.} - 2\Phi[D_c - D_d K_{eq.}]}{2D_c + D_d K_{eq.} + \Phi[D_c - D_d K_{eq.}]} \tag{6}$$

A detailed derivation of Maxwell's equation is given in this module's Appendix, which should be studied in order to work the problems at the end of this module. Equation 6 is rigorously valid only for a dilute suspension of randomly-distributed spheres, where the particles are of uniform size. For volume fractions, Φ, approaching zero, all other relationships derived for spheres in a continuum must give the same numerical value as Maxwell's equation. There are situations, however, where Equation 6 can also be applied to very concentrated suspensions ($\Phi \to 1$). This will be discussed later.

Let's consider for the moment the problem of the transport of a gas through a layer of uniform suspended spheres. According to Henry's law, the concentration is given by

$$C = \alpha p \tag{7}$$

where α is the solubility and p is the partial pressure. For the heterogeneous medium, the solubility is

$$\alpha_t = \alpha_c(1 - \Phi) + \alpha_d \Phi \tag{8}$$

The permeability P is defined as the product of diffusivity times solubility,

$$P = D\alpha \tag{9}$$

If Maxwell's equation is expressed in terms of permeabilities, for the effective permeability $\bar{P}(\bar{P} = \bar{D}\alpha_t)$,

$$\bar{P} = P_c \frac{P_d + 2P_c - 2\Phi[P_c - P_d]}{P_d + 2P_c + \Phi[P_c - P_d]} \tag{10}$$

Equation 10 has considerable use in the problem of oxygen transport in emulsions, dispersions and cellular suspensions.

Fricke (*4*) considered a suspension of randomly oriented ellipsoidal particles. His procedure was similar to that of Maxwell's, but instead of solving the potential problem for a single sphere, he considered the more general case of a spheroid and worked in confocal ellipsoidal coordinates. He obtained

$$\bar{D}K_t = D_c \frac{D_c + \Phi(TD_d K_{eq.} - D_c)}{D_c + \Phi D_c(T-1)} \tag{11}$$

where

$$T = \sum_{i=1}^{3} \frac{\frac{1}{3} D_c}{D_c - (D_c - D_d K_{eq.})I_i}$$

The three values of I_i involve elliptic integrals of the second kind and are purely numerical. These values depend only on the axial ratio of the dispersed particles. Fricke obtained solutions for particles which are spheroids. Here, two of the principal axes of the particle are of equal length and therefore, only one elliptical integral needs to be evaluated, since

$$I_1 + I_2 + I_3 = 1$$

and

$$I_2 = I_3$$

If the three principal axes of the spheroid are designated by a, b, and c (where $b = c$), then the values of I_3 as a function of the ratio of b/a are as shown in Table 1. For the case of spheres $a = b = c$, all I_i are equal to 1/3, and Fricke's equation reduces to Maxwell's. Note that when the ratio of b/a varies between its extremes, the particle shape changes from rod-shape to disk-shape. Fricke's equation is rigorously valid only for dilute dispersions of spheroids, and the volume fraction range in which the equation is accurate should be the same as for Maxwell's equation.

Bruggeman (*5*) derived a relationship for suspensions of spheres in which he attempted to take into account the interaction of neighboring particles. He obtained

$$\frac{\bar{D}K_t - D_d K_{eq.}}{(\bar{D}K_t)^{1/3}} = \frac{(1-\Phi)\ (D_c - D_d K_{eq.})}{(D_c)^{1/3}} \tag{12}$$

Table 1. I_3 for Spheroids ($b = c$)*.

Shape	b/a	I_3
Rods	0	0.5000
Prolate Spheroids	1/6	0.4784
	1/4	0.4624
	1/2	0.4134
Spheres	1	0.3333
Oblate Spheroids	2	0.2383
	4	0.1477
	6	0.1077
Disks	∞	0

* Values from Meredith and Tobias (1962).

The expression was derived for dispersions of random particles in the concentrated range of volume fractions by approximations based on Maxwell's relationship. Bruggeman used an integral procedure of adding infinitesimal fractions of dispersed phase to the mixture and considering the surrounding medium (mixture) for the added particles to be the continuum. Maxwell's equation was used to derive the differential change of the effective permeability with a differential change in the volume fraction. Subsequent integration yielded Equation 12. If one considers the procedure, Bruggeman's equation is a good approximation when a large range of particle sizes is present in the dispersion and the particle size distribution is polymodal (i.e., the mixture contains a large number of volume fractions of spheres with a uniform particle size, but where each particle size is widely different from any other particle size).

These relationships for the effective diffusivity are classical ones for particles suspended in a continuum. The Fricke and Bruggeman equations are variations of the Maxwell approach, where either the effect of the shape of the particles or the particle interaction is included. Relations are also available that consider either particle orientation or arrangement.

The Maxwell and Fricke equations are rigorously valid only for dilute suspensions, i.e., $\Phi < 0.1$. The relationships can be written in the form

$$\frac{\bar{D}K_t}{D_c} = 1 + a\Phi + b\Phi^2 + c\Phi^3 + \cdots$$

The Maxwell and Fricke relationships are only exact to the order Φ. For higher powers of Φ the equations are in error. For volume fractions near zero the terms higher than Φ are negligible. Considerable theoretical work is still in progress to evaluate the terms for the higher powers of Φ which are due to particle interactions (*6*).

There are situations, however, where the equations of Maxwell and Fricke can be applied to the highest volume fractions ($\Phi \rightarrow 1$). From a physical viewpoint, it is fortuitous that the relationships satisfy the following criteria:

when $\Phi = 0$, then $\bar{D} = D_c$

when $\Phi = 1$, then $\bar{D} = D_d$

The equations give the correct values at the extremes. For two-phase media where the diffusion properties of the two media are within one order of magnitude, the differences between the relationships and other theoretical developments that include particle interactions, are small. For example, it is obvious in Equation 10, that if the gas permeabilities in dispersed and continuous phase are equal, i.e., $P_d = P_c$, Maxwell's equation can be applied to all volume fractions. When P_d is within one order of magnitude from P_c, Maxwell's equation gives a reasonable prediction compared with more complex relations. In addition, Cole, et al (*7*) has shown experimentally that Maxwell's equation holds to volume fractions of 0.9 for suspensions with impermeable particles. The particles were either spherical, cubical, or hexagonal. As a consequence, Maxwell's equation should give good predictions for "spherical-like" particles at concentrated volume fractions, as long as P_d is not greater than P_c by one order of magnitude [Stroeve et al. (*8*)]. Spherical-like particles are defined here as those particles in which the ratio of minimum particle length to maximum particle length is near one. For systems where P_d for the spherical particles is much greater than P_c ($K_{eq.}D_d \gg D_c$), Maxwell's equation is no longer applicable to concentrated suspensions. It has been found experimentally that in this case, the data scatter between the predictions of Bruggeman and Maxwell.

It should be noted here, that for suspensions with monodisperse spheres which are nondeformable, the maximum volume fraction is $\pi/6 \approx 0.52$ for a cubic array of spheres, and $\pi/3\sqrt{2} \approx 0.74$ for a hexagonal packing of spheres. For suspensions with deformable spheres (such as an emulsion), spherical-like particles, or a wide particle size distribution, a higher volume fraction may be attained.

LAMINATED MEDIA

The orientation of the layers of the two-phase laminated media, relative to the direction of the flux, is important. Usually the layers are either normal or parallel to the flux. These two arrangements also give a set of limits (upper and lower bounds) for the effective diffusivity of two-phase media with one phase dispersed in the other. When the layers are parallel with the direction of the flux, the solution of the effective diffusivity is simply

$$\bar{D}K_t = D_c \frac{(1-\Phi)D_c + \Phi D_d K_{eq.}}{D_c} \tag{13}$$

Here the phases have been given subscripts *c* and *d*, in order to calculate the total flux with Equation 5, i.e., Φ is the volume fraction of phase *d*. When the layers are normal to the direction of flux, there is a series arrangement and the relationship is

$$\bar{D}K_t = D_c \frac{D_d K_{eq.}}{(1-\Phi)D_d K_{eq.} + \Phi D_c} \tag{14}$$

which is the relationship for the summation of resistances. The parallel and series arrangement solutions are exact and can be applied to all volume fractions.

DISCUSSION

The few theoretical relations presented here and in the Appendix are designed to give students an insight into the development of theoretical relations for effective parameters for heterogeneous media. Our concern here is with mass transfer in heterogeneous media; however, if appropriate modifications are made, the relationships are also applicable to other transport phenomena. The literature cited in this module are appropriate sources to find solutions for particle shapes other than the one discussed here. Further, recent theoretical work has also been done on mass transfer with chemical reactions in either the dispersed phase, the continuous phase, or both. In this case, the problem is not only to determine an effective diffusion coefficient, but in the case of irreversible chemical reaction, also an effective kinetic parameter for the heterogeneous medium as a whole.

This module has focused on the classical but limited equations of Maxwell, Fricke, and Bruggeman. A more recent approach to model transport processes in heterogeneous media is the method of volume averaging (*9–11*). In this method the local conservation equations in each phase are volume-averaged. Subject to certain constraints, the equations for each phase can be combined into a one-equation model for transport in heterogeneous media when the principle of local equilibrium is invoked. The resulting one-equation model contains the expressions for the effective physical and kinetic parameters. The method of volume averaging has several advantages over the classical techniques. First, the volume-averaged equations contain an attractive degree of generality because they require no specific geometry of the heterogeneous system for their development. In addition, the constraints that the one-equation model must satisfy are readily identified. Finally, the method of volume averaging can be used, with uniform rigor, to attack all types of multiphase transport phenomena. A recent exposition of the method has been given by Carbonell and Whitaker (*12*).

Note: This module was accepted for publication in January 1978.

LITERATURE CITED

1. Barrer, R. M., "Diffusion and Permeation in Heterogeneous Media," in "Diffusion in Polymers," J. Crank and G. S. Park, eds., p. 165–217, Academic Press (1968).
2. Meredith, R. E. and C. W. Tobias, "Conduction in Heterogeneous Systems," *Adv. Electrochem. and Electrochem. Engr.*, *2*, 15–47 (1962).
3. Maxwell, J. C., "A Treatise on Electricity and Magnetism," Vol. 1, 3rd Ed., p. 440, Clarendon Press, London (1904).
4. Fricke, H., "A Mathematical Treatment of the Electric Conductivity and Capacity of Disperse Systems," *Physical Rev.*, *24*, 575 (1924).
5. Bruggeman, D., "Berechnung Verschiedener Physikalischer Konstanten von Heterogenen Substanzen," *Ann. Physic.*, *5*, 636 (1935).
6. Batchelor, G. K., "Transport Properties of Two-Phase Materials with Random Structure," *Ann. Rev. Fluid Mechan.*, *6*, 227 (1974).
7. Cole, K. S., et al, "Electrical Analogues for Tissues," *Explt. Neurol.*, *24*, 459 (1969).
8. Stroeve, P. et al, "Steady-State Diffusion of Oxygen in Red Blood Cell and Model Suspensions," *AIChE J.*, *22*, 1133 (1976).
9. Anderson, T. B., and R. Jackson, "A Fluid Mechanical Description of Fluidized Beds," *Ind. Eng. Chem. Fundam.*, *6*, 527 (1967).
10. Slattery, J. C., "Flow of Viscoelastic Fluids Through Porous Media," *AIChE J. 13,* 1066 (1967).
11. Whitaker, S., "Diffusion and Dispersion in Porous Media," *AIChE J., 13,* 420 (1967).
12. Carbonell, R. G., and S. Whitaker, "Heat and Mass Transport in Porous Media," in "Mechanics of Fluid in Porous Media," J. Bear and Y. Corapcioglu, Eds., Martinus Nijhoff, Brussels (1983).

APPENDIX: MAXWELL PROBLEM

Single Sphere in a Continuum

Let us derive Maxwell's equation for gas transport. Consider a single sphere, of radius a and permeability P_d, immersed in an infinite liquid medium of permeability P_c. A constant mass flux N_A is imposed at large distances from the sphere, as shown in Figure 2. Note that maintaining either the mass flux or the driving force gradient $\Delta p/L$ (or potential gradient) constant is equivalent since at large distances from the sphere ($\theta = 0$),

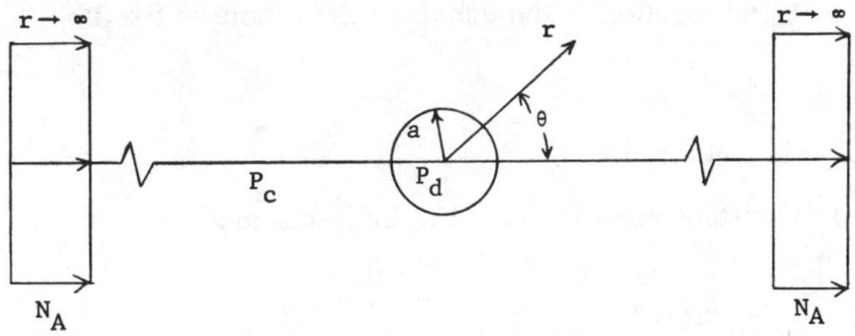

Figure 2. Single sphere in an infinite medium.

$$N_A = P_c \frac{\Delta p}{\Delta r}$$

The mass conservation equations for A inside and outside the sphere are:

outside sphere

$$P_c\left\{\frac{1}{r^2}\frac{\partial}{\partial r}\left(r^2\frac{\partial p_A}{\partial r}\right) + \frac{1}{r^2 \sin\theta}\frac{\partial}{\partial\theta}\left(\sin\theta\frac{\partial p_A}{\partial\theta}\right)\right\} = 0 \tag{A.1}$$

inside sphere

$$P_d\left\{\frac{1}{r^2}\frac{\partial}{\partial r}\left(\frac{r^2 \partial p'_A}{\partial r}\right) + \frac{1}{r^2 \sin\theta}\frac{\partial}{\partial\theta}\left(\sin\theta\frac{\partial p'_A}{\partial\theta}\right)\right\} = 0 \tag{A.2}$$

In deriving Equations A.1 and A.2, it was assumed that Henry's Law is applicable,

$$C_A = \alpha_A p_A$$

which is a reasonable assumption for sparingly soluble gases such as oxygen. The permeability is therefore defined as the diffusivity D_A times the solubility α_A of A in the medium. The boundary conditions are as follows:

1. Continuity of the potential p_A at the phase boundary:

$$p_A = p'_A \quad \text{at } r = a, \text{ all } \theta \tag{A.3}$$

2. Continuity of the mass flux of A at the phase boundary:

$$P_c\frac{\partial p_A}{\partial r} = P_d\frac{\partial p'_A}{\partial r} \quad \text{at } r = a, \text{ all } \theta \tag{A.4}$$

3. Concentration at the center of the sphere is fixed:

$$p'_A = p^o_A \quad \text{at } r=0 \tag{A.5}$$

4. Constant mass flux of A at large distances from the sphere:

$$P_c \frac{\partial p_A}{\partial r} = -N_A \cos\theta \quad \text{as } r \to \infty, \text{ all } \theta \tag{A.6}$$

The last boundary condition also imposes the condition of symmetry on the solution of Equation A.1 and A.2. The solution of A.1 and A.2 is attained through the method of the separation of variables by letting (for the inside potential or the outside potential)

$$p = R(r)\Theta(\theta)$$

The separated equations are of the form

$$\frac{\partial}{\partial r}\left(r^2 \frac{\partial R}{\partial r}\right) - n(n+1)R = 0$$

$$\frac{1}{\sin\theta}\frac{\partial}{\partial\theta}\left(\sin\theta \frac{\partial\Theta}{\partial\theta}\right) + n(n+1)\Theta = 0$$

The general solution of the differential equation $R(r)$ is

$$R(r) = A_n r^n + B_n r^{-n-1}$$

and the general solution for $\Theta(\theta)$ is

$$\Theta = E_m P^m_n(\cos\theta) + D_M Q^m_n(\cos\theta)$$

The functions P^m_n $(\cos\theta)$ and Q^m_n $(\cos\theta)$ are the associated Legendre functions of the first and second kind, respectively.

Application of the boundary conditions Equations A.3 through A.6 yields:

outside the sphere

$$(p_A - p^o_A) = -\left(\frac{N_A}{P_c}\right) r\cos\theta \cdot \left\{1 - \left(\frac{a}{r}\right)^3 \left[\frac{P_d - P_c}{P_d + 2P_c}\right]\right\} \tag{A.7}$$

inside the sphere

$$p'_A - p^o_A = -\left(\frac{N_A}{P_c}\right) r\cos\theta \left[\frac{3P_c}{2P_c + P_d}\right] \tag{A.8}$$

Outside the sphere, the partial pressure of A or potential field of A, $(p_A - p^o_A)$, is given by the imposed potential gradient,

$$\frac{N_A}{P_c} = \frac{\Delta p_A}{\Delta r} \tag{A.9}$$

at $r \to \infty$, times $r\cos\theta$ and a term accounting for the disturbance arising from the presence of the sphere. The relative effect of this disturbance term decays with the cube of the distance from the center of the sphere.

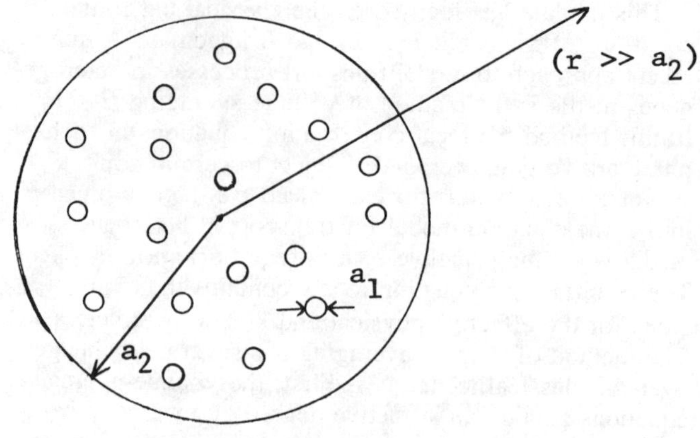

Figure 3. Spherical suspension of spheres.

Effective Permeability for an Ensemble of Spheres

Consider a dilute suspension of n spheres, with permeability P_d and radius a_1, located in a spherical volume of the continuous medium, with radius a_2, which itself is located in an infinite bath of the continuous medium which has a constant mass flux (or potential gradient) imposed on it (see Figure 3). The suspension is assumed to be sufficiently dilute that the interactions between spheres may be neglected. Under these conditions, at large distances from the center of the spherical cluster, the disturbance in the potential field arising from the presence of the n spheres is simply the sum of the disturbances resulting from each sphere individually. From Equation A.7, it follows that

$$p_A - p^o_A = -\left(\frac{N_A}{P_c}\right) r\cos\theta \left\{1 - \frac{n a_1^3}{r^3}\left[\frac{P_d - P_c}{P_d + 2P_c}\right]\right\} \tag{A.10}$$

The disturbance at such a point may also be written in terms of an effective permeability of the spherical cluster as a whole, $\bar{P}$,

$$p_A - p^o_A = -\left(\frac{N_A}{P_c}\right) r\cos\theta \cdot \left\{1 - \frac{a_2^3}{r^3}\left[\left(\frac{\bar{P} - P_c}{\bar{P} + 2P_c}\right)\right]\right\} \tag{A.11}$$

By equating the two expressions (A.10 and A.11) and using the volume fraction Φ of the dispersed phase,

$$\Phi = \frac{n a_1^3}{a_2^3} \tag{A.12}$$

one arrives at Maxwell's expression,

$$\frac{\bar{P}}{P_c} = \frac{P_d + 2P_c - 2\Phi[P_c - P_d]}{P_d + 2P_c + \Phi[P_c - P_d]} \tag{A.13}$$

STUDY PROBLEMS

1. What is Maxwell's equation for the case of impermeable (non-conducting) spheres suspended in a continuum in terms of $\bar{D}K_t/D_c = \square$?
2. For small volume fractions what is the relationship in Study Problem 1 for Φ approaching zero?
3. If the spheres are permeable, but $K_{eq.}$ approaches zero, is the relationship in Study Problem 1 still applicable?
4. Express $K_{eq.}$ and K_t in terms of the gas solubility coefficients.
5. Derive Equation 14 for a three layer laminate (i.e., *cdc*) with total thickness L and layer thicknesses of Δ_c and Δ_d.
6. Derive Equation 14 for a laminate of total thickness L consisting of a large number (i) of alternating layers (i.e., *cdcdcd* ····· *cdc*) with layer thicknesses Δ_c and Δ_d.
7. Study Problems 5 and 6 give as a final result, Equation 14. If there are either 3 or i layers, the flux through L, the laminates, will be the same, as long as L and Φ for the two laminates are equal. What can you say about the concentration profiles through these two laminates?
8. If the diffusivities D_c and D_d are the same in a system such as B in Figure 1, is $\bar{D}$ then equal to D_c or D_d?
9. What is the electrical conductivity $\bar{\chi}$ for a suspension of spheres according to Maxwell's equation?
10. Show that Fricke's equation reduces to Maxwell's for $a/b = 1$.

HOMEWORK PROBLEMS

1. Plot Maxwell's, Fricke's and Bruggeman's equation on a graph for the case that $D_d K_{eq.}/D_c = 5$. Use as axes $\bar{D}K_t/D_c$ and Φ. For Fricke's equation, consider particles with an axial ratio $b/a = 4$. What can you say about the three relationships? What can you say about the Maxwell and Bruggeman relationship at low Φ? What can you say about Fricke's relationship at low Φ (with $b/a = 4$)?
2. Plot also the same relationships for $D_d K_{eq.}/D_c = 0$ (i.e., non-permeable or non-conducting particles). Where are the differences between the relationships greatest?
3. What are the boundary conditions in the Maxwell problem in the Appendix, if there is an interfacial resistance R_m at each fluid-particle interface?
4. For transport in porous solid material, such as catalyst particles, the experimental effective diffusion coefficient is often correlated with the porosity ϵ and the tortuosity factor τ,

$$D_{eff.} = \frac{D_c \epsilon}{\tau}$$

The tortuosity gives a measure of the increase in length and the possible constrictions in the diffusion paths within the continuous non-solid phase (*c*) of the porous solid. What is the prediction of τ from the model of Bruggeman? The solid phase is impermeable to solute transfer. Note that the steady-state flux is given by (slab geometry, thickness L)

$$J = D_{eff.} \frac{\Delta C_c}{L}$$

From heterogeneous media theory, the steady state flux is

$$J = \bar{D}K_t \frac{\Delta C_c}{L}$$

Module C6.6

Advancing Front Theory

R. Srinivasan

Air Products and Chemicals Inc.
Allentown, PA 18105

and

Pieter Stroeve

University of California
Davis, CA 95616

OBJECTIVES

At the completion of this module, the student should be able to:

1. Define the factors that influence the transport of material into reactive fluids flowing in conduits such as: length of exchanger, reactive capacity of the fluid, and penetration of the saturated reaction front into the fluid.
2. Understand the effect of chemical reaction upon the transport rate.
3. Be able to predict the necessary length of a conduit in order to accomplish a certain conversion of reactants.
4. Design artificial oxygenators utilizing advancing front theory.

PREREQUISITE MATHEMATICAL SKILLS

Be able to:

1. Solve partial differential equations.
2. Dimensionalize equations.

PREREQUISITE ENGINEERING AND SCIENCE SKILLS

1. Understand basic mass conservation equations.
2. Be familiar with mixing cup concentrations.
3. Understand mass transfer coefficients and Sherwood numbers.
4. Studied the Graetz and Lévêque solutions for heat and mass transfer in conduits.
5. Had a basic introduction to the effect of chemical reaction on transport phenomena.

INTRODUCTION

What is the common denominator in the following phenomena: freezing of a lake, regeneration of a coked reformer catalyst, tarnishing of a metal surface, impregnation of supports with catalytic active materials, consolidation of clay layers in the soil, oxygenation of fatigued muscles or venous blood? All these phenomena involve the diffusion of mass, heat or some other scalar quantity, influenced by the presence of a moving boundary. The location of the moving boundary depends on the amount of diffusing quantity brought to it. The problem concerning the freezing of a lake was first studied by Stefan (*1*), hence, these moving boundary problems are sometimes called "Stefan Problems."

The presence of a moving boundary introduces a nonlinearity into the diffusion equation. Consequently, only a few moving boundary problems have exact analytical solutions. Monographs and reviews detail analytical and numerical techniques that have been employed to solve these problems (*2, 3*). This module will be restricted to mass transfer and specifically, to an integral technique which makes use of the "advancing front" concept. The assumptions and the method of this technique will be clear from the examples presented. The following examples are for some biological problems; however, they are applicable to other reactive systems, as well.

UNSTEADY-STATE DIFFUSION IN A STATIONARY REACTIVE MEDIUM

Consider a planar semi-infinite body that contains a reactant B, which is suddenly exposed at its surface to a constant concentration of species A. Suppose that the species A reacts infinitely fast with B, to form a product P, i.e., $A + B \rightarrow P$. The penetration of A into the semi-infinite body would be slow, compared to the situation where no reactant B is present. At time t, the concentration profile of A in the body could be shown in Figure 1. The species A has penetrated a distance η into the semi-infinite body. Since the kinetics of the reaction is infinitely fast, A and B cannot co-exist so that, in the region $0 \leq x \leq \eta$, all B is converted to P. In the region $\eta < x < \infty$, no A is present and the concentration of B

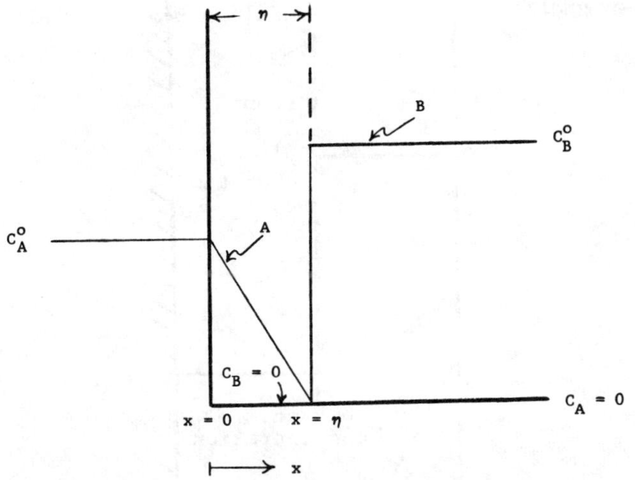

Figure 1. Diffusion of a species *A* into a stationary reactive medium initially containing the immobile reactant *B*. *A* reacts infinitely fast with *B*. The profiles are shown at some time *t* which is the time that the medium has been exposed to concentrations C_A^o.

equals its initial value, C_B^o. The distance η grows with time; it advances like a sharp front. Two examples that are closely representative of this system can be given.

Diffusion of Oxygen into a System Containing Lactic Acid

Hill and Marrack (*4*) studied the rate of diffusion of oxygen into fatigued muscle. Oxygen is depleted and lactic acid is produced inside a muscle, which is subjected to continuous strenuous exercise. When the muscle is brought to rest and oxygen is introduced suddenly around it, the process of recovery to a normal state begins at once. Oxygen diffuses into the muscle and reacts with lactic acid. The kinetics of this reaction is very fast, so that oxygen and lactic acid react in an extremely thin layer (i.e., at a front) which slowly sweeps forward through the tissue until recovery is completed. Hill and Marrack assumed that the reaction was infinitely fast and that the diffusion of lactic acid is so slow compared to diffusion of oxygen that, essentially, oxygen alone diffuses. The equation describing the diffusion of oxygen, is then

$$\frac{\partial C}{\partial t} = D_c \frac{\partial^2 C}{\partial x^2} \qquad 0 \le x \le \eta \tag{1}$$

Initial condition:

At $t = 0$, there is an oxygen debt λ moles oxygen/cc of tissue.

Boundary conditions:

At $x = 0$,

$C = C_o$, where $t \ge 0$

At $x = \eta$,

$$-D_c\left(\frac{dC}{dx}\right) = \lambda \frac{d\eta}{dt}$$

The last boundary condition equates the rate of arrival of oxygen at the reaction zone with the rate at which the oxygen debt is satisfied. Hill and Marrack introduced a variable

$$z = \frac{x}{2\sqrt{D_c t}}$$

and arrived at

$$C_o = \lambda\sqrt{\pi}\xi e^{-\xi^2}\left[\frac{2}{\sqrt{\pi}}\int_0^{\xi} e^{-\theta^2}\, d\theta\right] \tag{2}$$

where

$$\xi = \eta/2\sqrt{D_c t}$$

and where θ is a dummy variable. The term in the brackets is the error function, erf ξ. From Equation 2, the penetration depth of the front, η, can be calculated. By introducing an approximation, a simpler solution for η can be derived as follows. Assume the oxygen front to advance so slowly into the tissue that a steady-state concentration profile is set up behind it. In this case, Equation 1 becomes

$$\frac{dC}{dt} = 0 = D_c \frac{d^2C}{dx^2} \tag{3}$$

from which it follows that

$$C = C_o(1 - x/\eta) \tag{4}$$

At the advancing front, as before,

$$\lambda \frac{d\eta}{dt} = -D_c\left(\frac{dC}{dx}\right)_{x=\eta}$$

$$= D_c C_o/\eta$$

One arrives at the ordinary differential equation,

$$\eta \frac{d\eta}{dt} = D_c \frac{C_o}{\lambda}$$

from which,

$$\int_o^{\eta} \eta\, d\eta = \int_o^t D_c \frac{C_o}{\lambda}\, dt$$

or

$$\eta = \sqrt{\frac{2D_c C_o t}{\lambda}} \tag{5}$$

The advancing front depth predicted by Equation 5 is a good approximation for the value of η obtained from Equation 2, as long as C_o is small and λ is large. Note that Equation 5 predicts that the depth of the advancing front varies with the square root of time. The total amount of oxygen reacted with lactic acid per unit cross-sectional area of tissue is

$$V = \eta\lambda = \sqrt{2D_c C_o \lambda t} \tag{6}$$

Hill and Marrack also considered the case that lactic acid is not immobilized in the tissue, but can diffuse from the bulk to the reaction zone. For $x > \eta$, a mass balance equation similar to Equation 1 is written for lactic acid. The boundary condition at the advancing front is modified, such that the ratio of the fluxes of oxygen and lactic acid at the front equals the stoichiometric ratio in which they react. This gives an analytical solution similar to Equation 2. For the special case in which the advancing front moves very slowly into the medium, so that steady state profiles of oxygen and lactic acid are set up behind and ahead of the front, an approximate but explicit expression for η is obtained.

$$\eta = D_c \frac{C_o}{\lambda} \sqrt{\frac{\pi t}{D_L}} \tag{7}$$

where D_L is the diffusivity of lactic acid.

The solutions by Hill and Marrack are the same as those derived later for absorption of a gas into a liquid with rapid chemical reaction (*5*). These solutions form special cases in Danckwerts' general study of unsteady state diffusion with a moving boundary (*6*).

Oxygenation of a Stationary Film of Hemoglobin or Blood

Engineers have made significant contributions in the study of the diffusion of oxygen in blood. An understanding of the diffusion process is of importance in designing artificial oxygenators for respiratory support. Oxygen reacts rapidly and reversibly with hemoglobin present in red blood cells. In artificial oxygenators, the blood is contacted with almost pure oxygen at atmospheric pressure. The oxygen partial pressure in the gas is so high that the reaction is forced far to the right and can be modelled as

$$O_2 + Hb \rightarrow HbO_2$$

where Hb denotes hemoglobin and HbO_2, oxyhemoglobin. In practice, venous blood has a certain initial concentration of oxyhemoglobin. The fraction of the total hemoglobin content that exists as oxyhemoglobin is called the fractional saturation of hemoglobin, *S*.

$$S = \frac{C_{Hb}}{C_{Hb} + C_{HbO_2}} = \frac{C_{Hb}}{c_T}$$

where c_T is the total hemoglobin concentration. Consider the diffusion of oxygen into a film of blood, as illustrated in Figure 2. For infinitely fast reaction, the total oxygen balance is(*7*):

$$\frac{\partial}{\partial t}(C + 4c_T S) = D_c \frac{\partial^2 C}{\partial x^2} + 4c_T D_{Hb} \frac{\partial^2 S}{\partial x^2} \tag{8}$$

C is the concentration of free oxygen and $4\ c_T S$ is the

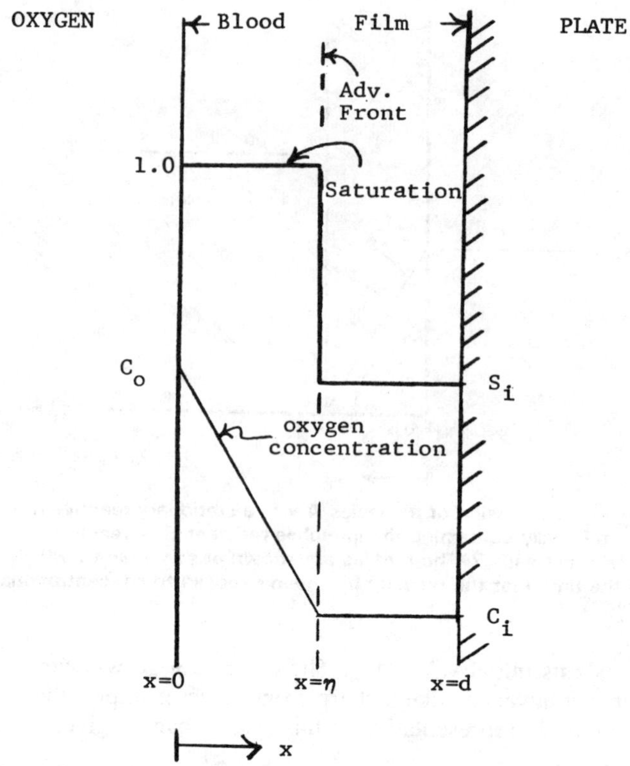

Figure 2. Penetration of oxygen into a film of unsaturated hemoglobin solution (or blood) at initial saturation S_i.

concentration of chemically combined oxygen (the factor 4 arises because four oxygen molecules combine with one hemoglobin molecule).

Initial Condition:

At $t < 0$, $C = C_i$, all x

Boundary Conditions:

At $x = 0$, $C = C_o$, $t \geq 0$

At $x = d$, $\partial C/\partial x = 0$

The fractional saturation curve relating *S* to *C* is nonlinear, and for low partial pressures of oxygen in the surrounding gas, Equation 8 normally has to be solved numerically. For the case of high oxygen partial pressures, assuming that hemoglobin does not diffuse, Marx et al (*8*) conducted an analog computer study and concluded that the advancing front can be approximated by

$$\eta = \sqrt{\frac{2D_c(C_o - C_i)t}{4c_T(1 - S_i)}}$$

Note that the term $4\ c_T(1 - S_i)$ is the remaining oxygen capacity of the blood and is equivalent to λ of the previous example. The study of Marx et al, supports the concept that there is a sharp boundary between fully saturated hemoglobin ($S = 1$) and unsaturated hemoglobin ($S = S_i$) and that there exists a linear oxygen concentration gradient in the saturated layer (Figure 2). Coupling these approximations with the assumption that $\eta = \alpha t^{1/2}$, where α is an as yet undetermined constant, the oxygen balance at the front can be written as

$$\int_o^t D_c \frac{(C_o - C_i)}{\eta}\, dt = 4c_T(1 - S_i)\eta + \frac{(C_o - C_i)}{2}\eta \tag{9}$$

where S_i is the initial saturation. The second term on the right-hand side represents physically dissolved oxygen. Substitution of the assumed form of η into Equation 9 gives

$$\int_o^t D_c \frac{(C_o - C_i)}{\alpha t^{1/2}}\, dt = 4c_T(1 - S_i)\alpha t^{1/2} + \frac{(C_o - C_i)}{2}\alpha t^{1/2} \tag{10}$$

The indicated integration leads to

$$\alpha = \sqrt{\frac{2D_c(C_o - C_i)}{4c_T(1 - S_i) + (C_o - C_i)/2}} \tag{11}$$

so that,

$$\eta = \sqrt{\frac{2D_c(C_o - C_i)t}{4c_T(1 - S_i) + (C_o - C_i)/2}} \tag{12}$$

The instantaneous average saturation of the blood on the plate is

$$S(t) = \frac{\eta}{d}(1 - S_i) + S_i$$

$$= \left\{ \frac{2D_c(C_o - C_i)t}{d^2\left[4c_T(1 - S_i) + \frac{1}{2}(C_o - C_i)\right]} \right\}^{1/2} (1 - S_i) + S_i \tag{13}$$

Spaan (*7*) has shown that the solution of Equation 8 for $S(t)$ with $D_{Hb} = 0$ is predicted well by Equation 13 over a wide range of values for C_o, c_T and S_i.

STEADY-STATE DIFFUSION IN A FLOWING REACTIVE FLUID

Now let us consider the situation of blood, or a hemoglobin solution, flowing along a flat plate, in a film of thickness d. At the free surface ($x = 0$), oxygen diffuses into the blood, as shown in Figure 3. Assuming laminar flow and a fully developed velocity profile, the oxygen balance is given as (*7–10*):

$$v_z \frac{\partial}{\partial z}(4c_T S + C) = D_c \frac{\partial^2 C}{\partial x^2} + 4c_T D_{Hb} \frac{\partial^2 S}{\partial x^2} \tag{14}$$

The boundary conditions are

$$\text{At } x = d,\ \frac{\partial C}{\partial x} = 0 \tag{15a}$$

$$\text{At } x = 0,\ C = C_o,\ z \geq 0 \tag{15b}$$

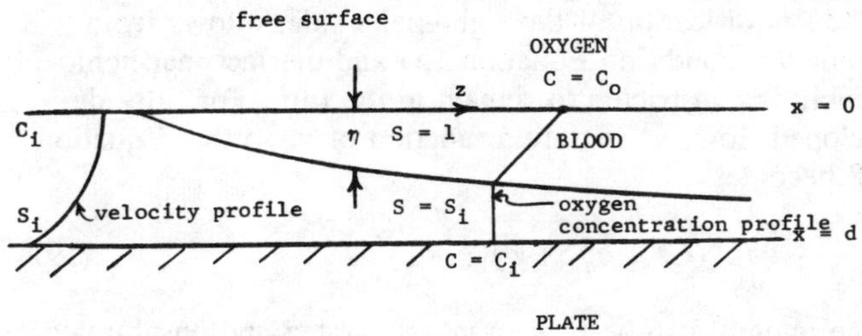

Figure 3. Penetration of oxygen into a film of unsaturated hemoglobin or blood, of initial saturation S_i, flowing over a flat plate.

$$\text{At } z < 0,\ S = S_i,\ C = C_i \tag{15c}$$

For convenience, Equation 14 is made dimensionless,

$$v^* \frac{\partial}{\partial z^*}(C^* + c_T^* S) = \frac{\partial^2 C^*}{\partial y^2} + D^* \frac{\partial^2 S}{\partial y^2} \tag{16}$$

where

$$C^* = C/C_o$$

$$C_i^* = C_i/C_o$$

$$c_T^* = 4c_T/C_o$$

$$y = x/d$$

$$D^* = c_T^* D_{Hb}/D_c$$

$$v^* = v_z/\bar{v} \tag{17}$$

and

$$z^* = zD_c/d^2\bar{v}$$

$\bar{v}$ is the average velocity of the film. A solution which takes into account the nonlinear saturation curve would have to be numerical. Again, however, the advancing front hypothesis can be used to obtain approximate solutions as was done by Lightfoot (*11*), Dorson (*10*) and Spaan (*7*). The assumptions in the application of the advancing front hypothesis have been summarized by Lautier, et al (*12*):

i) The flow of blood is Newtonian.
ii) The velocity profile is fully developed.
iii) Axial diffusion is negligible compared to radial diffusion.
iv) Metabolic consumption of oxygen by blood is negligible.
v) The reaction of oxygen with hemoglobin is instantaneous.
vi) The oxygen concentration profile in the saturated layer is linear.

Now, Equation 16 can be integrated across the film thickness.

$$\int_o^1 v^* \frac{\partial}{\partial z^*}(C^* + c_T^* S)\, dy = \int_o^1 \left(\frac{\partial^2 C^*}{\partial y^2} + D^* \frac{\partial^2 S}{\partial y^2}\right) dy$$

$$= -\left(\frac{\partial C^*}{\partial y}\right)\Bigg|_{y=0} \tag{18}$$

The simplification of the right-hand side follows from boundary condition Equation 15a and the fact that hemoglobin is constricted to remain in the film. For fully developed flow, v^* is only a function of y, so that Equation 18 becomes

$$\frac{d}{dz^*}\int_o^1 v^*(C^*+c_T^*S)\,dy=-\left.\frac{\partial C^*}{\partial y}\right|_{y=0} \tag{19}$$

The integral can be broken into saturated and unsaturated regions.

$$\frac{d}{dz^*}\left[\int_o^p v^*(C^*+c_T^*S)\,dy+\int_p^1 v^*(C^*+c_T^*S)\,dy\right]=-\left.\frac{\partial C^*}{\partial y}\right|_{y=0} \tag{20}$$

Here p is the dimensionless thickness of the saturated layer, $p = \eta/d$ and

$$y<p,\ S=1 \text{ and } C^*=1-\left(\frac{1-C_i^*}{p}\right)y$$

$$y>p,\ S=S_i \text{ and } C^*=C_i^*$$

After substituting the conditions, which are based on the assumption of a linear concentration profile, the right-hand side of Equation 20 becomes

$$-\left.\frac{\partial C^*}{\partial y}\right|_{y=0}=\frac{1-C_i^*}{p} \tag{21}$$

For laminar flow over a flat plate,

$$v^*=\frac{3}{2}(1-y^2) \tag{22}$$

Introduction of these relationships into Equation 20 gives

$$\frac{d}{dz^*}\left[\int_o^p \frac{3}{2}(1-y^2)\left(1-\frac{(1-C_i^*)}{p}y+c_T^*\right)dy+\int_p^1\frac{3}{2}(1-y^2)\,(C_i^*+c_T^*S_i)\,dy\right]=\frac{1-C_i^*}{p} \tag{23}$$

or

$$\frac{3}{2}\frac{d}{dz^*}\left[\left(p-\frac{p^3}{3}\right)(c_T^*+1-C_i^*-c_T^*S_i)-\left(\frac{p}{2}-\frac{p^3}{4}\right)(1-C_i^*)+\frac{2}{3}(C_i^*+c_T^*S_i)\right]=\frac{1-C_i^*}{p} \tag{24}$$

Carrying out the indicated differentiation,

$$\frac{3}{2}\left\{(1-p^2)[c_T^*(1-S_i)+(1-C_i^*)]-\left[\left(\frac{1}{2}-\frac{3p^2}{4}\right)(1-C_i^*)\right]\right\}\frac{dp}{dz^*}=\frac{1-C_i^*}{p} \tag{25}$$

Multiplying by p and integrating, this equation becomes

$$z^*=\frac{3}{2}\left[\frac{p^2}{4}(1+Q)-\frac{p^4}{8}\left(Q+\frac{1}{2}\right)\right] \tag{26}$$

where

$$Q=\frac{(1-S_i)\ 4c_T}{\frac{1}{2}(C_o-C_i)} \tag{27}$$

The average saturation $\bar{S}$ attained in dimensionless length z^* is

$$\begin{aligned}\bar{S}&=\int_o^p v^*S\,dy+\int_p^1 v^*S\,dy\\&=\frac{3}{2}\int_o^p(1-y^2)\,dy+\frac{3}{2}S_i\int_p^1(1-y^2)\,dy\\&=S_i+\frac{3}{2}(1-S_i)\left(p-\frac{p^3}{3}\right)\end{aligned} \tag{28}$$

For the case of a film of blood flowing between two parallel membranes with oxygen diffusing into both sides of the liquid film of thickness $2d$, the dimensionless length of the exchanger and the average saturation are given by Spaan (*7*) as

$$z^*=3\left[-p^4\left(\frac{1}{32}+\frac{Q}{16}\right)+p^3\left(\frac{1}{9}+\frac{Q}{6}\right)\right] \tag{29}$$

$$\bar{S}=S_i+(1-S_i)\left(-\frac{1}{2}p^3+\frac{3}{2}p^2\right) \tag{30}$$

With Equations 29 and 30, an oxygenator with a parallel plate configuration can be designed.

GENERALIZED ADVANCING FRONT THEORY

By modifying the boundary condition of Equation 15b to include a membrane resistance and by keeping the definition of the velocity v^* variable, Oomens and Spaan (*13*) generalized the advancing front theory to be applicable to a variety of flow situations. The result is an expression for the dimensionless exchanger length in terms of three integrals, one representing chemical reaction, one representing the membrane resistance and the third, representing physically dissolved oxygen. By substituting the appropriate velocity profile in this expression, oxygen uptake can be predicted for each different flow situation.

CORRELATIONS

The engineer is faced with the questions: How long will the oxygenator have to be to give a particular saturation of the blood, and how do different oxygenators compare in mass transfer efficiency for saturating blood? The answers can be obtained in a variety of ways. For instance, one can plot

$$\frac{\bar{S}-S_i}{1-S_i} \text{ versus } \left(\frac{2}{Q}\right) z^*$$

or

$$\frac{\bar{S}-S_i}{1-S_i} \text{ versus } \left(\frac{2}{1+Q}\right) z^*$$

as was done by Lightfoot (*11*) or

$$\bar{S} \text{ versus } z^*$$

as was done by Spaan (*7*). It is recognized that ($1/z^*$) is equivalent to the Graetz number.

$$N_{Gz}=\frac{1}{z^*}=\frac{d^2\bar{v}}{zD}=\frac{d}{z}\cdot N_{Re}\cdot P_r$$

Terms such as

$$\frac{(1+Q)}{2z^*} \text{ or } \frac{Q}{2z^*}$$

are known as modified Graetz numbers (N^*_{Gz}).

Since this is mass transfer in a conduit, the oxygenation process can also be characterized in terms of a plot of the Sherwood number versus the dimensionless axial distance, or a modified Graetz number. In a parallel membrane device, the mass transfer rate per unit area,

$$= -D\left.\frac{dC}{dx}\right|_{x=0}$$

$$=\frac{D}{\xi}(C_o-C_i)$$

$$=\frac{D}{p\cdot d}(C_o-C_i) \tag{31}$$

Also, the mass transfer rate per unit area is

$$=k_m(C_o-C_B)$$

where k_m is the mass transfer coefficient and C_B is the mixing cup oxygen concentration.

$$C_B=\int_o^1 Cv^*\, dy$$

$$=\frac{3}{2}\int_o^1 C(2y-y^2)\, dy$$

$$C_B=\frac{3}{2}\int_o^p\left[C_o-(C_o-C_i)\frac{y}{p}\right][2y-y^2]\, dy$$

$$+\frac{3}{2}\int_p^1 C_i[2y-y^2]\, dy$$

The velocity profile v^* is for flow between parallel plates. Carrying out the integration, gives

$$C_B=C_i+(C_o-C_i)\ (p^2/2-p^3/8)$$

$\therefore$ Mass transfer rate per unit area $=k_m(C_o-C_B)$

$$=k_m(C_o-C_i)\left[1-\frac{p^2}{2}+\frac{p^3}{8}\right] \tag{32}$$

From Equations 31 and 32, the local fluid-side Sherwood number is

$$N_{Sh_z}=k_m\cdot\frac{2d}{D}=\frac{2}{\left(p-\frac{p^3}{2}+\frac{p^4}{8}\right)} \tag{33}$$

Since p is related to z^* and Q by Equation 29, the local Sherwood number can be expressed as

$$N_{Sh_z}=f_1(z^*,\ Q)$$

or

$$N_{Sh_z}=f_2(N_{Gz},\ Q) \tag{34}$$

or

$$N_{Sh_z}=f_3(N^*_{Gz})$$

with f_1, f_2, f_3 being different functions. Figure 4 is a plot of N_{Sh_z} versus z^* for the different values of Q in the parallel membrane situation. Included in this figure is a plot of N_{Sh_z} versus z^* for the nonreactive case obtained from Colton et al (*14*). The local Sherwood number for the reactive flow case is considerably higher than that for the nonreactive situation, since chemical reaction enhances the oxygen transfer rate. Note using nondimensionalization has reduced a complex problem to a relationship of three dimensionless numbers z^*, Q and N_{Sh_z} (*15*).

Another approach was discussed by Curtis and Eberhart (*16*). They noted that a plot of the terminal fractional saturation increase, $(\bar{S}-S_i)/(1-S_i)$, versus the inlet resistance, R, is an excellent performance pre-

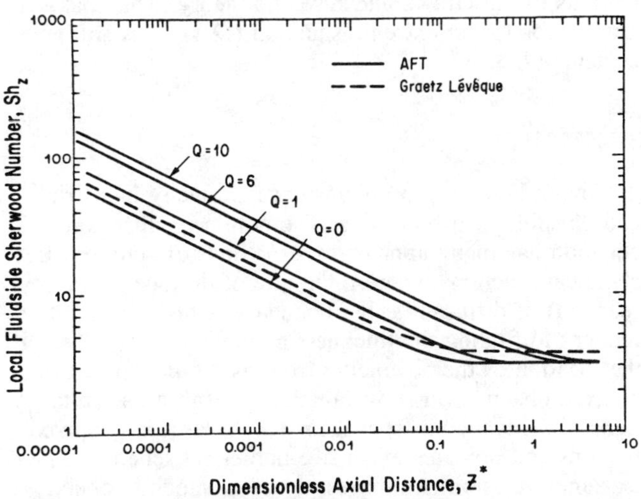

Figure 4. Local fluid-side Sherwood number as a function of the dimensionless axial distance and the reaction parameter *Q*.

dictor of oxygenators. The inlet resistance is defined as

$$R = \frac{1}{Q} \frac{1}{(\text{flow rate})}$$

It turns out that R is proportional to the modified Graetz number. By plotting either the fractional saturation increase or the Sherwood number versus z^* (or a Graetz number), at a fixed inlet saturation S_i and a fixed Q, the performance of different oxygenators can be compared. The plots can be based on the advancing front theory or experimental data. Apart from being useful in comparing oxygenators, such analyses can also determine the validity of the advancing front theory by comparison with experimental data.

LIMITATIONS OF THE ADVANCING FRONT THEORY (AFT)

The critical requirement for the applicability of the advancing front theory is that the time constant for movement of the reacting front be much smaller than the time constant for the diffusion process. Kutchai (*17*) pointed out that the blood layer should be thicker than 50μ to satisfy this requirement. Looking at Q, one sees that

$$Q \cong (\text{unbound oxygen})/(\text{undissolved oxygen})$$

or

$$Q \cong \frac{[\text{oxygen uptake by chemical combination}]}{[\text{dissolved oxygen concentration difference}]}$$

As Q decreases (e.g. due to a decrease in hematocrit of blood and consequently, hemoglobin concentration), the ability of blood to combine chemically with oxygen decreases. Since AFT is based on the assumption that this ability is large enough to result in the slow movement of the reaction front, it follows that for low values of Q, AFT predictions are in error. However, from Figure 4, the predictions of AFT theory for $Q = 0$, i.e. the nonreactive case, are still to within 15 percent of the predictions of the Graetz and Lévêque theory. This maximum error for an extreme situation ($Q = 0$) is still acceptable (*15*).

SUMMARY

Moving boundary problems occur in many industrial and physiological processes. The nonlinear diffusion equation has many approximate methods of solution. For chemical reactions wherein the rate of disappearance (by reaction) of diffusant at the boundary is fast, the solution reduces to finding the thickness of the boundary layer or the location of the advancing front as a function of time or axial distance (in flow situations). Within its limitations, the advancing front theory provides good approximations and obviates expensive numerical schemes when extreme precision is unnecessary. This module looked into physiological examples, but the theory can be applied to other reactive systems.

Note: This module was accepted for publication November 1978.

LITERATURE CITED

1. Stefan, J., "Ueber die Theorie der Eisbildung, insbesondere uber die Eisbildung im Polarmeere," *Ann. Physik und chemie, 42*(3), 269–286 (1891).
2. Ockendon, J. R. and W. R. Hodgkins, eds., "Moving Boundary Problems in Heat Flow and Diffusion," Clarendon Press, Oxford (1975).
3. Bankoff, S. G., "Heat Conduction or Diffusion with Change of Phase," *Adv. Chem. Engr., 5,* 75 (1964).
4. Hill, A. V. and P. E. Marrack, "Diffusion of Oxygen and Lactic Acid Through Tissues," *Proc. Roy. Soc., London, Ser. B, 104,* 39 (1928).
5. Sherwood, T. K. and R. L. Pigford, "Absorption and Extraction," p. 333, McGraw-Hill, New York (1952).
6. Danckwerts, P. V., "Gas-Liquid Reactions," McGraw-Hill Book Co., New York (1970).
7. Spaan, J. A. E., "Transfer of Oxygen into Hemoglobin Solution," *Pflügers Arch., 342,* 289 (1973).
8. Marx, T. I., et al, "Diffusion of Oxygen into a Film of Whole Blood," *J. Appl. Physiol., 15,* 1123 (1960).
9. Buckles, R. G., et al, "An Analysis of Oxygen Absorption in a Tubular Membrane Oxygenator," *AIChE J., 14,* 703 (1968).
10. Dorson, Jr., W. J., "Oxygenation of Blood for Clinical Applications," in "Blood Oxygenation," E. D. Hershey, ed., Plenum Press, New York (1970).
11. Lightfoot, E. N., "Low Order Approximations for Membrane Blood Oxygenators," *AIChE J., 14,* 669 (1968).
12. Lautier, A., et al, "The Use of Mathematical and Physical Models for Optimization of Oxygenators," *Trans. Am. Soc. Artif. Int. Org.,* XVII, 303 (1971).
13. Oomens, J. M. M., and J. A. E. Spaan, "A Generalized Advancing Front Model Describing the Oxygen Transfer in Blood," *Adv. Exptl. Med. and Biol., 75,* 233 (1976).
14. Colton, C. K., et al, "Laminar Flow Mass Transfer in a Flat Duct with Permeable Walls," *AIChE J., 17,* 773 (1971).
15. Stroeve, P. and R. Srinivasan, "An Approximate Solution for the Graetz and Lévêque Problems from the Advancing Front Theory," *AIChE J., 26, 136* (1980).
16. Curtis, R. M. and R. C. Eberhart, "Normalization of Oxygen Transfer Data in Membrane Oxygenation," *Trans. Amer. Soc. Artif. Int. Org., 20,* 210 (1974).
17. Kutchai, H., "Wider Applicability for Hill's Advancing Front Theory of Oxygen Uptake," *J. Appl. Physiol., 31,* 302 (1971).

STUDY PROBLEMS

1. Sketch qualitatively the increase of the average saturation with the dimensionless length for oxygen gas contacting blood flowing over a flat plate.
2. For Study Problem 1, plot the saturation profile into the layer at a given distance z.
3. For a stagnant layer of reactant B, where B is highly mobile in the layer, show qualitatively the concentration profile for reactants A and B. How does the advancing front profile compare to the profile for B?
4. An artificial oxygenator, composed of semipermeable parallel membranes, has a length z. Draw the expected change in the average saturation between outlet and inlet blood as a function of the average velocity.
5. If the velocity of the blood becomes extremely slow, what is the outlet average saturation?
6. It is known that the reaction of oxygen with hemoglo-

bin is reversible. Why does one assume in an artificial oxygenator that this reaction is irreversible?

7. The enhancement factor E is defined as the increase in the mass transport rate due to chemical reaction compared to the nonreactive case. For a given z^*, how can this enhancement factor be obtained from Figure 4?

HOMEWORK PROBLEMS

1. In Figure 4, for the nonreactive case, the Lévêque solution yields for the Sherwood number a solution of the form

$$N_{Sh_z} \propto z^{*-1/3}$$

This solution is applicable only to small z^*. The Lévêque solution therefore represents the straight line portion of the dashed line which has a slope of $-1/3$ on the log–log plot. For small z^*, the advancing front solution also yields a Lévêque type solution. What is it?

2. Based on the solution obtained in the Homework Problem 1, what is the enhancement factor E as discussed in Study Problem 7?

3. What would be an appropriate dimensionless plot to bring all reactive curves of the Sherwood number (as derived from the advancing front theory) into one single curve?

4. In a manner analogous to the derivation of Equations 26 and 28, derive Equations 29 and 30 for the case of flow between two flat membranes, with oxygen diffusing into both sides of the liquid film of thickness $2d$.

APPENDIX

Solutions to the Study Problems

Solutions to the Homework Problems are available as a separate reprint from the AIChE Educational Services Dept., 345 East 47th St., New York, NY 10017.

SOLUTIONS TO THE STUDY PROBLEMS

1. For any bubble wall velocity the kinetic energy of the liquid would be infinite for both the cylindrical and plane bubbles. Hence growth of such bubbles is impossible.

2. (1) No the flow is irrotational.
   (2) The radial stress may be expressed in terms of wall velocity.
   (3) Use the momentum solution for this estimate.

3. Neglect successively convective and transient terms.

4. Reasonably good agreement obtained.

SOLUTIONS TO THE STUDY PROBLEMS

1. "Dispersion" is a general term which designates mixing. While such mixing at the microscopic level always occurs by molecular diffusion, on a macroscopic scale, it is aided by convective transport due to a non-uniform velocity profile or turbulent eddy transport. The axial dispersion of a solute in flow through a conduit is the result of a complex interaction between the mechanisms of convective and diffusive transport.

2. Axial dispersion due to convection is caused by the fact that in a non-uniform velocity field, elements of solute at different transverse locations translate down the conduit at different velocities -- the effect of such differences in velocity is to smear out an initially compact solute cloud. Thus, velocity variations across the cross-section are essential for causing axial dispersion. In a uniform velocity field, the solute cloud translates as a whole at a uniform velocity and, therefore, there is no spreading due to convection.

3. There is no need for a parabolic velocity profile to cause convective dispersion. Any non-uniform velocity field will be adequate.

4. Taylor's principal contribution was to formulate the basis for a theoretical description of the dispersion process and to derive an expression from theory for the dispersion coefficient k -- he also showed that his model of the dispersion process agreed with the experimental observations of Griffiths as well as his own.

5. For simplifying his treatment, Taylor ignored the axial molecular diffusion term in his theoretical formulation. Aris in a more general treatment, using the method of moments, showed that the effect of axial molecular diffusion is additive -- that is, to include the effect, all one needs to do is to add the value of the diffusivity to the dispersion coefficient.

6. Diffusion in the axial direction helps axial spreading or axial dispersion. Diffusion in the transverse direction inhibits axial spreading.

SOLUTIONS TO THE STUDY PROBLEMS

1. Diffusional coupling effects would be most important in terms of transport across tissue membranes; individual cells cannot sustain significant bulk flow in or out for any length of time. The kidney and intestines (small intestine in particular) are relevant organs. Water is by far the most important more permeable component.

2. Even if $D_B = D_A$, it is still possible to get countergradient transport if distribution coefficients k_A and k_B are different enough to generate sufficient bulk flow $K = n_A + n_B$.

3. If $\omega_{A1} = \omega_{A2}$, Equations (6) and (7) don't apply because $K = 0$ in this case.

4. The minimums in the D_B/D_A curves of Figure 2 exist because of two competing effects. As ω_{A2} decreases from 0.3 for $D_B/D_A > 1$, a non-zero total flux K results. This total flux increases as ω_{A2} decreases toward zero. However, as ω_{A2} decreases toward zero there is less A present to be carried by this bulk flow even though the flow itself is of increasing absolute magnitude.

SOLUTIONS TO THE STUDY PROBLEMS

1. Steady state and equilibrium both require no change with time, i.e. $\partial/\partial t = 0$. However, equilibrium in addition requires that no gradients exist. Thus, for a chemical reaction at equilibrium, there is no net production or disappearance of any chemical species. For the facilitated diffusion analysis in terms of D_{eff}, we have equilibrium for reaction but steady state for the diffusion.

2. a. Facilitation F would not double if $p_{①}$ were doubled from 100mmHg to 200mmHg. The reason is that the Hb is already about 97% saturated as HbO_2 at 100mmHg.

   b. If $p_{②}$ were fixed at 150 mmHg for $p_{②} < p_{①}$, facilitation F would be essentially zero. The reason is that $S \cong 1$ for all pressures greater than about 120mmHg. Thus dS/dC_{O_2} in Equation (11) would be zero throughout the solution.

3. a. For collisional transport it would be expected that F would increase continuously as T increases because of faster collision rates.

   b. For collision it would be expected that F would depend on heme group concentration squared since facilitation would depend on $Hb^* + HbO_2 \rightleftharpoons Hb^*O_2 + Hb$. The dependencies of a) and b) are not observed.

4. A substance of high molecular weight can be dilute in molar concentration and non-dilute in mass concentration. Fifteen grams of Hb per 100 gm of water is about 13 weight % Hb. In terms of mole fraction, however, $x_{Hb} = 1.58 \times 10^{-4}$ based on a Hb molecular weight of 17,000.

SOLUTIONS TO THE STUDY PROBLEMS

1. $\dfrac{\bar{D}\,K_t}{D_c} = \dfrac{2(1-\Phi)}{2+\Phi}$

2. $\dfrac{\bar{D}\,K_t}{D_c} = 1 - 1.5\,\Phi$

   if $\Phi \rightarrow 0$

3. Yes

4. $K_{eq} = \dfrac{\alpha_d}{\alpha_c}$

   $K_t = (1-\Phi) + \Phi\,\dfrac{\alpha_d}{\alpha_c}$

5. The derivation of equation 14 is simple as it is a problem of addition of resistances. At some point of the derivation one must define $\Delta_d = \Phi$.

6. The derivation of equation 14 is similar as that for problem 5 except that it is algebraically more involved. As an alternative problem equation 13 can be derived for layers parallel to the direction of flux.

7. The total concentration profiles are different. In each individual c layer the slope of dC_c/dx is the same; similarly in each d layer the slope of dC_d/dx is the same.

8. No. Only if K_{eq} is equal to one.

9. Same as Maxwell's equation for gas transport but with $\bar{\kappa} = \bar{P}$, $\kappa_c = P_c$ and $\kappa_d = P_d$.

10. $I_3 = 1/3$.

SOLUTIONS TO THE STUDY PROBLEMS

1.

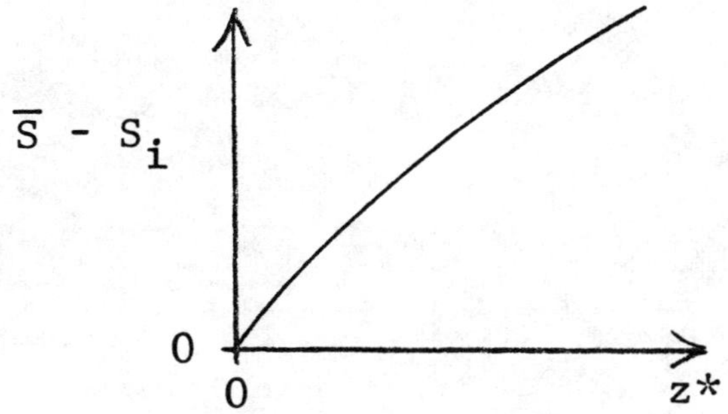

2.

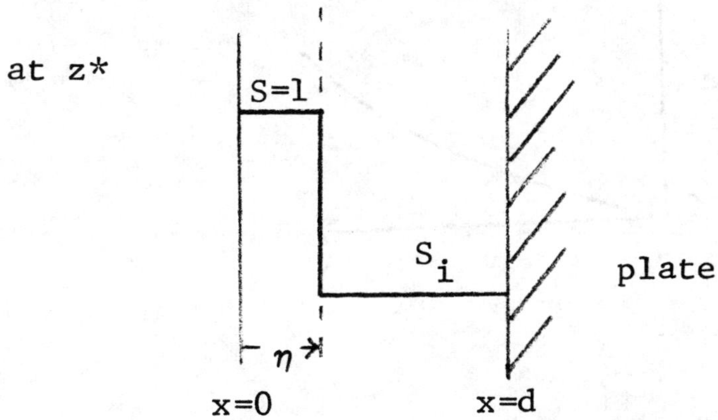

3.

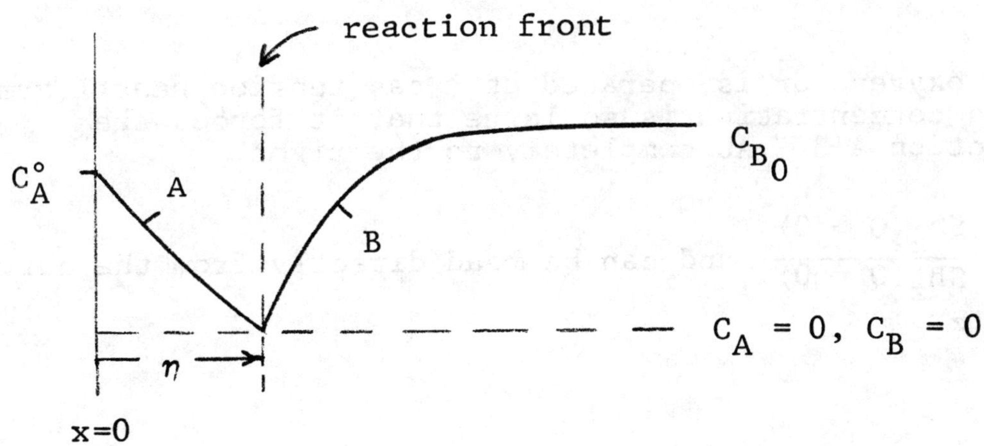

4. The faster the flow the smaller $\overline{S} - S_i$. However the total mass transfer rate

$$N_A \cong v_{AV} \times (\overline{S} - S_i) \times \text{flow area}$$

increases with increasing flow

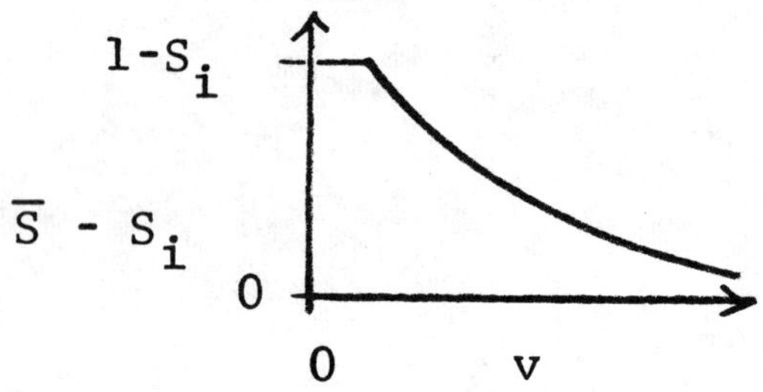

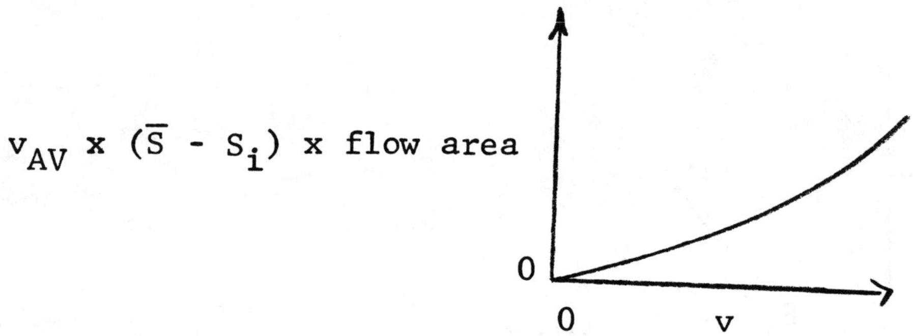

5. $\overline{S} - S_i = 1 - S_i$

so $\overline{S} = 1$ as shown above

6. The oxygenator is operated at a gas tension near 1 atm. This concentration is so large that it forces the reaction A+B ⇄ AB completely to the right.

7. $E = \dfrac{Sh_z(Q > 0)}{Sh_z(Q = 0)}$ and can be read directly from the curves at any z*.